COURAGEOUSLY
BROKEN

COURAGEOUSLY BROKEN

A memoir about overcoming adversity and conquering the battle scars of life

D.A. MICHAELS

GSM Books

This memoir reflects the author's life faithfully rendered to the
best of her ability. Some names and identifying details have been
changed to protect the privacy of others.

D.A. Michaels
facebook/d.a.michaels
www.courageously-broken.com

First edition: November 2020

ISBNs: 978-1-0879-0744-4 (hardcover)
978-1-7353413-1-6 (paperback)
978-1-0879-0745-1 (ebook)

GSM

Dedication

This labor of love is dedicated to
my incredibly bright, kind and beautiful daughter.
You are the Greatest Blessing a mother
could ever hope for and I am eternally grateful for the gift of you.
I am proud of you beyond words for your strong leadership skills
and steadfast morale compass. Stay just the way you are.
I love you to the moon, stars,
to infinity and beyond the heavens

~ Mom

Contents

Foreword

So often, when military or law enforcement personnel share stories of the daily hardships, perseverance, and struggles that they confront on a daily basis, they are the stories of men. Rarely has the story ever been told from a female perspective. Donna Michaels, who served both in the US Navy and as a police officer vividly describes her life in a wide open, personal and heartfelt fashion.

I had the pleasure of serving with Donna in the Republic of Panama during the Pablo Escobar and Manual Noriega hunt and the Central and South America drug wars (1989 – 1993). Reading her story made me feel as if I was right back in Central America, operating in the jungles, conducting missions from the riverine and coastal patrol boats and doing all we could do to reduce the drug trafficking coming into and out of that very corrupt part of the world.

Although she developed a real sense of patriotism and desire to serve her community from some of her close relatives, she was also subjected to a great deal of emotional and physical abuse from some of these same family members, co-workers and her close friends. She painfully describes the dysfunctional and abusive relationship she had with her father, which more than likely, created the very rocky path she had later in life with the boys and men with whom she had relationships.

Donna's life, filled with the many great experiences she had while traveling the world and eventually finding peace is one that I very much relate to and one that most anyone who served in the military or in the law enforcement community can relate to. Her story is also a fascinating and informative read for those who did not serve as it portrays how life in these communities, especially as a female, so often plays out.

CWO3 Don D. Mann
Navy SEAL Team 6 (retired)
NY Times Best Selling Author

NO MATTER HOW MUCH IT HURTS,
HOW DARK IT GETS, OR
HOW FAR YOU FALL,
YOU ARE NEVER OUT OF THE FIGHT.

~ MARCUS LUTTRELL

One

⚜

Hindsight is 20/20

Last night, like many nights, I awoke from a dream that terrified me. I was drenched in sweat, heart pounding, gasping for breath and uncertain of where I was. Then I realized, I was home, safe in my bed and reminded myself that was all behind me. After a few deep breaths and the comfort of my service dog, Saint, I was able to fall back asleep.

There was a time when that dream and the real-life incidents that precipitated it would have sent me into a tailspin that lasted for days. Crying hysterically while suffering from exhaustion as I pushed myself through the motions of life.

The truth is I never imagined I would be where I am today. I'm approaching the end of a successful law enforcement career, I'm the mother of an amazing daughter who's nothing short of miraculous. I own the home and car of my dreams and I have somehow become comfortable in my own skin. On the surface, my life looks quite normal. I'm safer and more secure than I ever dreamed possible, yet I struggle to believe that and I wonder if I ever will.

The road to get here certainly wasn't easy and it probably never will be. More than once, I've wanted to run away and disappear, giving in to the black cloud that seems to follow me wherever I go.

I never thought I'd be sharing my story with the world or anyone outside my small circle of trust. In the quest of trying to make sense of things, I began a journal in an attempt to take a look at my life and make sense of how I got to where I am today. I wanted to overcome adversities and extraordinary traumatic events that nearly killed me three years ago. I've been told so many times over the years how strong I am, but I've never felt strong. In fact, I've spent much of my life feeling quite broken.

As the chapters of my life fell onto paper, I realized that sharing my story might encourage others who have struggled with their own battles. Perhaps I can inspire others to understand they aren't alone and that they too can learn to embrace their scars.

My hope is to give a shining light to anyone who may be going through or who has gone through similar tragedies and are asking themselves "Why?" or worse, telling themselves they can't take the pain anymore or feel as though they are a burden to their loved ones. These struggling thoughts and feelings are indescribable and until they happen to you or someone you love, you genuinely have no idea how extremely difficult it is to have them.

So this is me, a fifty year old woman, with PTSD, who served her country in the military abroad and her community at home as a law enforcement officer, pondering *What's next?* If by some chance you find yourself thinking *PTSD isn't real, it's something only the weak suffer from*, I will tell you I used to believe the exact same thing.. Until it caught up to me.

As I sit here nearing the end of my law enforcement career reflecting on my life and seeing where I am today, never in a million years did I see myself exploring and contemplating the next chapter. Perhaps before we can move forward, we need to look back and use the gift of hindsight to understand how we got to where we are today.

Two

❧

There's No Such Thing as Perfect

Cocoa Beach, Florida 1962-1983

Everyone wants a perfect family, but as we grow, we realize there is no such thing. Even those we believe to be the most perfect have ugly secrets kept out of the public's eye. While I didn't grow up in the public eye or come from a family who did, my mother did her best to portray our family as perfect. She was a product of the 50's and image was everything. She didn't believe in complaining or airing out our dirty laundry, even to me, her only child.

My mom grew up between Fort Lauderdale, Florida and Detroit, Michigan, the daughter of a WWII Navy Seabee and a nightclub singer who couldn't make up their minds where they wanted to live. She went to 14 schools growing up since they moved about once a year. Her parents married just before Grandpa was sent off to WWII. My uncle was born nine months later and Mom just after the war ended. Mom worshiped her father even though he wasn't a touchy feely kinda man. She resented her mother a great deal for her alcoholic ways and repeatedly embarrassing her. Mom recalls being the woman of the house, responsible for cooking, cleaning and working in addition to going to school. Her only outlet was her love to dance. Growing up in the 50's, school dances were held regularly and she lived for them. She would win every dance competition and never struggled for a date. While getting a date for a dance was easy, finding a steady boyfriend wasn't because she was a "good girl". If I was a betting woman, my bet is that Grandpa scared the shit out of the boys so they wouldn't ask her out.

She met my father when she was 16 and he was 34. He was working on the NASA salvage boats with Grandpa and his brother, a former Navy diver. She married him as soon as she turned 17 with her parents' permission. A decision she said she would later regret if it hadn't been for me. She was pregnant seven times and lost every pregnancy except me, so I grew up an only child. After fifty years, March 10th is still a difficult day for her, as that is the birth date of a baby boy born full term with the umbilical cord wrapped several times around his neck. He didn't survive. Later she would be responsible for typing the autopsy notes on her own child as there was no one else who could. To everyone's surprise, she got pregnant right away and just over a year later, I was born.

Mom worked hard at the local hospital in the pathology department. She was the administrative coordinator between five forensic pathologists. She dealt with cops, lawyers, grieving family members and doctors from all over the hospital. She described it as challenging and stressful but loved every minute of it.

She was old fashioned, a virgin until she married, preached that girls don't fight or curse, and despite the fact that she could no longer go to Communion because she married a man who was divorced, she remained a devout Catholic. She was a peace keeper and got along with everyone. Nonetheless, if she believed in something, really believed in it, she could ruffle feathers standing up for what she felt was right. My father called her a passive aggressive once and she smacked the shit out of him. Something that shocked us both. He actually laughed because it was so out of character for her.

As a young girl, I believed my family was just like any other. A mom, a dad and an incredible dog. A German shepherd. Barney was so much more than a dog to all of us, he was a member of the family and a sibling to me. We loved him with all our hearts. Barney was police K9 trained and my parents knew leaving me home alone with him at a young age would not be a problem. God help anyone that even looked at me wrong. He was so sweet, but threaten his pack and he would tear them to shreds. He once chased a phony utility guy up our telephone pole in the backyard and held him there until the police arrived. It turned out he was suspected of bugging people's phone lines to get their personal information.

Barney died at the age of 13 after hip dysplasia got the best of him. It was the first time I saw my father cry. I would dream about him for years and I could hear his collar jingle in the empty house as I would arrive home after school every day. It was as though he never left, but he had, and we missed him terribly.

My father was a technical writer at NASA. He grew up in a predominantly German neighborhood in Cincinnati, Ohio. The son of a raging alcoholic father and a mother with a vicious tongue. He never spoke of his father and we never really knew much of his childhood, but when he did speak of his father, it wasn't good. He told the story of how he was a young boy with pneumonia when his father threw him out into the snow to chop firewood. Later he would quit school to join the Navy just to get away. He said he was prone to getting in trouble and his uncle was the Chief of Police. His uncle told him he would be on his own the next time he got into trouble and he had covered for him for the last time, so he joined the Navy. WWII had just ended and Mom was born that same year; it was 1946.

While working at NASA, Dad got laid off a lot with each government contract that would expire and later I learned he only worked seven years of my parents' 30 year marriage. Yet, he was the boss and what he said went. Dad would do strange things like stay up all night on his HAM radio and much like his teenage years, he often got into trouble that his brother, now a homicide detective, would have to cover up for him. Neither of them had the greatest moral compass.

My worst young memory of my father was when he decided to get rid of my dachshund. As much as I loved Barney, he was dad's dog and way too big to put in my lap. I wanted a dog of my own that was more "pint sized". Unfortunately, my dachshund, Sebastian, had an attitude problem and would often try to display dominance over Barney. One day he decided it would be a good idea to pick a fight with Barney over a toy. In the course of this, he bit him on the lip which undoubtedly hurt. In the blink of an eye, before anyone could say "No!" Barney had Sebastian by the throat and was shaking him like a rag. After a week in ICU, Sebastian returned home without any vocal cords. The first thing he did upon coming home was growl at Barney. It was clear, he had not learned his lesson. I was about six years old and instead of sitting me down and explaining to me it was for the dogs own good and safety, Dad decided it would be a better idea to just grab the dog and drop him off in a random neighborhood in the rain. Like any child, I threw a fit and cried. All I remember as I tried to stop him from walking out that door with my dog, was him backhanding me into a wall. That was my first nosebleed, but not my last beating.

Before being old enough to be home alone and when dad was working or when my parents would go out at night, my father's mother would watch me. Grandma loved watching me as I was the youngest of all her grandchildren. My uncle had four children, dad had four from his previous marriage that he left behind, so Grandma relished her time with me and spoiled the crap outta me when she could. She rubbed a lot of people the wrong way because of her outspoken ways. Once I remember going out late with her on a Saturday night to go fishing, she LOVED to fish. As she arrived on the riverfront, a popular place for teenagers to "go parking," I remember her walking to each car, banging on the windows while yelling at them to "knock that shit off! I've got my granddaughter and we are going fishing!" I was too young to understand what they must have been doing, but even today looking back

I remember the embarrassment it caused me. Now I get a good laugh wondering what must have been going through those teenagers' heads with the crazy old lady beating on their window.

I went to Catholic School from kindergarten through seventh grade. Each year as we decorated for Christmas, I looked forward to one particular cherished ornament, a simple red angel. It was no masterpiece, that was for sure, but it held a great story that brought a smile to my face every year. As I sat in art class in Sister Benedict's classroom, all the children were carefully painting their angels. I was not, nor have I ever been the artistic type. So there I sat, arms folded with a scornful face. I wanted no part of this project. So Sister Benedict stood over me calling me out in front of the class demanding I paint this ornament for my mom to put on the tree. I looked up at her defiantly, grabbed the angel, and dunked it in the closest bucket of paint there was which happened to be bright red. As I dunked the angel and pulled it out, firmly setting on the newspaper in front of me, I said, "There! It's painted!" Sister didn't say a word, she merely walked away and explained to my mom later why my angel was a solid red. Mom was cool about it, turns out she hated art too growing up.

When I was in first grade, my father was out of work again, he was responsible for getting me ready and off to school. I rode the school bus home in the afternoons. One day I staggered up the driveway with my tattered uniform, shirt hanging out and one ponytail in, one out. My lunchbox was hanging open and I could barely put one foot in front of the other. My father greeted me asking what happened to me. To which I replied, "I had a bad day daddy." He asked, "Was Sister Helen mean to you honey?" I replied, "No, Sister Helen wasn't there today, we had a prostitute teacher." Needless to say, the Irish Nuns laughed until they cried when they heard what I said and in the many years that followed, they remembered me as the student who had the prostitute teacher.

The neighborhood I grew up in was made up of predominantly Italian Catholics and as a German, French and Native American descendant, I didn't realize I wasn't Italian. I always enjoyed going to all the Italian Catholic weddings. They were so much fun and I dreamed of one day being a bride in a big wedding just like all the ones I loved attending. The entire European culture fascinated me because of the little exposure I had as a child.

Western North Carolina 1983-1989

In 1983, I was 13 when my father decided we were moving to the remote mountains of North Carolina. A far cry from the sun and sand of Cocoa Beach, Florida. My mom left the hospital and the people she loved and I was pulled from the only school I'd ever known one year from graduating with my friends. I was less than happy about the move.

Dad had these grandiose plans of becoming a professional photographer. Unfortunately he lacked the people skills and motivation to get his business off the ground. Instead, his behavior worsened as he spent most of his time on the HAM radio and became obsessed with conspiracy theories. As usual, my mom was working her ass off trying to make ends meet and I just did my best to lay low. Sometimes it worked, sometimes it didn't, depending on his day.

Despite what my mother would tell you, I had wonderful grandparents. They also built a home in the remote mountains just down the hill from ours. My grandmother eventually got her drinking under control and while many would tell you she was a pain in the ass, I would argue and say she was my ally. Whenever things would get to be too much, Grandma was always there to listen and guide me and I was always grateful for that.

Grandpa was a man's man. Everyone loved him despite his grouchy demeanor. He could make anyone laugh and keep an absolute straight face. I grew up understanding why my mom idolized him the way she did. I was his favorite grandchild of nine. I'm not sure why, but he favored me until the day he died at the age of 88. He would take me to the airport when I was little just to watch the planes take off. Later as a teen he would bail me out of trouble when I'd wreck my ATV off the side of a mountain road. Not once did he ever rat me out to my parents.

He got a kick out of my adventurous side. As much as I hated boxing, I loved sitting with him to watch a match. He absolutely loved the sport.

As I grew up and began to spend time with other friends and their families, I realized my family was not like them. My friends didn't walk on eggshells like my Mother and I did. I started to see the difference between my parents and others. I began to realize my father wasn't a nice guy and began to resent him for how he treated my mother and me. Always controlling, everything was quid pro quo. He spied on us and no matter what we did, he would accuse us of the most bizarre things. He even went as far as tapping our phone lines at home to listen to our conversations. My mother found a wire that ran through the house to a recording device. While she was at work and I was at school, he would spend the day at home listening to our conversations from the night before.

I remember numerous occasions where I stayed home from school as he sat in his chair surrounded by all his notebooks, magazines, tape recorders and a 357 magnum revolver. He had days he would talk about being better off dead and told me he wouldn't be around when I got home from school. So feeling responsible for him, I would stay home and spend the day talking him out of killing himself. I was only sixteen.

Each and every man in my family served in the military. The vast majority were in the U.S. Navy. My grandfather, the WWII Veteran was a Seabee in Guam. They cleared the island of the Japanese and built what became a US Navy Base. He would tell stories of how they needed supplies, to include beer and liquor and how he would always find a way to score some for the troops. He made sure to take care of the officers or "cake eaters" as he liked to call them so they would turn a blind eye when the enlisted men wanted to have a little fun. He was known as the "Scavenger" because he knew how to be resourceful and get things other people needed or wanted. Growing up in a family of 19 brothers and sisters in northern Michigan during the Great Depression would do that I imagine if you wanted to eat and survive.

My father served in the Navy for two years between World War II and Korea. He did his time on an aircraft carrier and while he didn't excel in the Navy, he had great stories about traveling through Europe which I had always dreamed of doing because of my Catholic upbringing. The nuns at school were from Ireland and I loved hearing their Irish accents and learning Irish songs. We learned all about Rome and Vatican City. I was fascinated with how beautiful it all sounded and the way of life there.

In 1983, I remember watching the news as we did every night. I watched and learned of the horrific attack on the USMC Barracks in Beirut. It troubled me a great deal and I felt a deep desire to do something to help. It was then and there I decided I would join the Navy when I graduated from high school. I had no desire to go to college and all I really wanted to do was see the world. Joining the Navy seemed like a positive way to fulfill my dreams.

As the years went by and I began to get more freedom as a teenager, I spent more and more time with friends. I saw the difference between their fathers and mine and I realized getting away from home and away from the small town we lived in was my only hope.

At the age of 16, I told my parents my plans of joining the Navy. I remember telling them that I didn't want them to say no and I wanted them to understand why I was pursuing this path. I did not expect them to support my goals, but I hoped they would. To my surprise, without hesitation, they were elated. My mom told me she had wanted to do the same when she was young but her father forbade it. Back then women in the military were frowned upon. This was now 1986 and everyone was planning for college. I was a smart child, I just hated reading and school. I wanted to learn about the world by experiencing it, so this was my plan.

That same year, my father suffered a tragic accident on my ATV. He went off a cliff and suffered a serious injury to his leg. In addition, his brain suffered a lack of oxygen which took his paranoia to a whole new level. I was driving by this time and had gotten a job. I was rarely home between school and work. I spent as much time away as I could because home was just a really miserable place to be.

Romantically I struggled. I always seemed to be attracted to the bad boys and got my heart broken over and over again. I had never had a positive role model from my father so I had no idea how to determine what was acceptable behavior from a boy and what wasn't. This only drove my desires to leave that small town even more. Once I had a huge crush on a guy from the next town over. Dean barely knew I existed, I questioned if he even knew my name. Every girl in town wanted to go out with Dean, but he was the quiet type who stayed to himself and spent most of his time away from school working. Nonetheless, I talked about him A LOT! Later I heard that not only did my father go seek this boy out, but my grandfather joined him. Probably one of the few times the two of them were on the same page. Dean was a few years older than me and had a reputation of being a "ladies man" because of his good looks. With that said, Dad and Grandpa took it upon themselves to find this kid working at the lumber yard, pulled a knife on him and threatened that if he even looked at me, they would cut his dick off. Needless to say, when word of that got out, not a single boy had the guts to ask me out.

The first time I saw a recruiter at school, I approached him and told him who I was and what I was going to do. He blew me off and told me to come see him in a couple years when I was old enough. I hounded that recruiter with questions every month when he came for his visit. By my senior year, I was working three jobs, maintaining good grades and was still head strong in my goals. I wanted to be a journalist because I loved writing and reporting problems and events. He sat me down and unlike any other recruiter, Kelly told me he was concerned. The movie *Top Gun* had come out the previous year and as a result, kids were trying to join the Navy in epic proportions. Everyone believed they could join the Navy, become a fighter pilot and get into "dog fights" while getting the girl. Kelly told me the real Navy was not what I thought it was. It wasn't like Hollywood portrayed it and he was concerned I would be devastated when I learned that. It was then I finally had the opportunity to tell him about how I came from a long line of Navy men and grew up hearing stories of mischief and heroism. Hell, I even got to tell him a ship had been named after one of my uncles in WWII. I explained to him how I wanted to serve my country and do my part to protect the homeland. Lastly I told him about my home life and how I knew there was nothing in that small town for me. I needed a way out and the Navy was it. He finally got it, I was different. He came to my home for dinner one night and met with my parents. He agreed to help me and he never sugar coated a thing. He guided me to be as prepared as I could possibly have been for boot camp and the challenges that would face me. I think I had that one in a million recruiter. No one else I've ever met could say the same thing about theirs.

In 1987 the *USS Stark* was attacked by an Iraqi jet during the Iraq Iran war. This struck me to my core. I couldn't wait to join the Navy and get in the thick of things. During our high school senior trip to Washington DC, my class was scheduled for a Pentagon tour. I followed our Navy tour guides like a lost puppy. My friends teased me terribly but I didn't care. I wanted to soak up as much as I could because I was ready to go.

Finally the time came to enlist. I made several trips to the Military Entry Processing Station, or MEPS, before I finally was offered a rate I was comfortable with. I had my hopes on being a journalist but those billets were far and few between. I had scored high enough on the ASVAB test to do just about whatever I wished. Writing was my passion and the rating of Yeoman looked interesting. My father told me it was a good choice because it would open a lot of opportunities for tour duty choices. "Yeoman's are needed everywhere," he said. My recruiter agreed so that was that. I was going to be a Yeoman.

I got sworn into the Delayed Entry Program but had to wait before I could leave for boot-camp based on my A school slot. I graduated high school in May of 1988 and killed time with summer jobs and friends until I left for the Navy in February 1989.

Just prior to leaving for the Navy, I got into my first fist fight. I was raised that girls don't fight, or at least that is what my mom always preached to me. My father had a different perspective, he kept telling me to grab a stick and beat the hell out of whoever was bothering me.

It would be years before I would figure out the right answer was somewhere in the middle.

I had endured years of harassment by this one specific loud mouth girl, Alicia. She was jealous because I briefly dated an ex boyfriend of hers. It was a week before I was leaving for boot camp, I was 18, an adult. I pulled into a popular hangout to say my goodbyes when she approached me running her mouth as usual. It was then and there I decided I needed to stand up for myself because I wanted to leave knowing I could. We stood there trash talking for the longest time when suddenly it happened. She pushed me and I saw red. All I remember was picturing a bullseye, a literal bullseye on her face and I reared back my fist and hit it as hard as I could. The next thing I knew, I had her in a headlock and I was beating away, years of being picked on coming out all at once. Someone broke up the fight and as I stood there with the chest of my sundress ripped away, exposing my bra and a tiny scratch on my shoulder, I began to laugh at her. There she stood with a bloody nose, bloody lip and hair a mess. She asked me what I was laughing at and I told her the truth, "I'm laughing at you!" I thanked her for allowing me the chance to do what I wanted to do for so long. To which she replied, "Well if you think it's so funny, why don't you try again?" I shrugged my shoulders, took a step forward and WHAM! Right in the eye. At that she fell back and her friends walked her away. Later I heard she suffered a broken jaw and nose. She told the folks at the emergency room she had fallen down stairs. It was the first and only time I called home and asked for dad. I knew my mom would lose her mind. Dad told me to come home and we'd talk about it. We waited all night for the cops to show up, but they never did. Years later I ran into her while home on leave and she was kind to me. We talked and she admitted she deserved that ass beating for treating me the way she did all those years. My only regret was that I hadn't stood up for myself sooner. Maybe high school would have been easier had I not been perceived as weak? It was years later I learned everyone in that town was laying bets on how long I'd last in the Navy. No one thought I'd even make it through Boot Camp.

1974 - My protector and me

Three

⚜

Boot Camp

Orlando, Florida February, 1989

My high school in North Carolina was very small. There were only 74 students in my graduating class and plenty were just like me with no college plans but a strong desire to serve our country. Thirteen of us joined, only two of us were girls. Charlene and I were polar opposites from each other because we came from extremely different backgrounds. She lived with several siblings in a small trailer on the side of a mountain and she was the epitome of a tomboy. I was viewed as the "rich city girl" even though my family was far from rich. I never really understood Charlene's purpose for joining the military as we were never very close and we didn't confide in each other. In hindsight, I think she struggled with her identity and instead of figuring that out, she joined the Navy to prove something. Whether it was to herself or everyone else, I'll never know. She was rough around the edges and could be pretty abrasive. She came from a dirt poor family and maybe she thought the Navy was a good way to just get out of her environment and pursue a better life.

I, on the other hand, embraced my femininity. I could ride an ATV, drive a 4X4 Jeep through the mud, get dirty and hang with the guys all day and still find pleasure in dressing up and putting on a pair of high heels later that night. My mom made me promise when I joined the Navy that I would remain a lady. This meant no tattoos, no cursing, and absolutely no sleeping around. Well, by the time I came home on my first leave, I had already screwed up two out of three.

The night before I left for boot camp, my parents drove me to Charlotte to go through MEPS. I remember dinner being very somber. We were all nervous but stayed positive hoping this was going to be a good decision. It was later I found out my Mom cried the entire four hour drive home. All I can remember is nausea and nerves. I don't think I slept a wink that night and getting up at 4 a.m. was brutal on all of us to get me where I needed to be. This was just the first of many many very early mornings.

I spent the day at MEPS where everyone goes through the process of getting vaccinations, physicals, answering personal questions, learning where to go, how to follow military etiquette, and a million other things I've long forgotten. I mostly remember being scared like everyone else. There were very few females and I learned I would be spending the night alone in a hotel because my flight for Basic Training in Orlando, Florida didn't leave until o'dark thirty the next morning.

I felt like a piece of meat. Every guy there that night wanted to get lucky one last time before he wasn't allowed to even look at a woman for the next several months. I remember having dinner and literally pushing a guy away as I locked myself in my hotel room. I hadn't worked my tail off and gotten that far only to screw things up. Unlike men, the quickest way for a woman to get booted out of the Navy was, and probably still is, getting pregnant. God knows I didn't need that. Plus, I was focused on making something of myself and becoming a mother was at the bottom of my priority list.

The next morning we got up and flew to Orlando. There we were picked up by a bus and driven to Naval Training Center Orlando. We were assigned a Company, mine was K049. Funny how I can remember things like that but not what I had for breakfast this morning. Anyway, my company, which consisted of all females, was housed on the second floor of a three story building. The floors above us and below us were companies of all males. The rare times we saw each other we dared not look at each other or there was hell to pay. My company was always getting cycled which meant we were forced to do repetitive physical exercise, earning us the nickname "Cycle Queens." Pushups, bodybuilders, jumping jacks for eternity, situps, anything and everything the company commanders could come up with to make us miserable. I think it's because we had some girls in our company who didn't take the time to get in shape before arriving, so we all paid for it.

My bunk-mate had a really difficult name to pronounce, Hodkiewicz, and mine was no picnic either. The first day they went through roll call, they got to my bunk-mate first. After several failed attempts at pronouncing her name, they decided on "Alphabet." It was really difficult not to laugh. Next on the roll was me. The company commander had no clue how to pronounce my name, so she showed the other company commander the clip board and said, "Who the fuck is this?" and that was that, for the next eight weeks, my name was "Hudafuck'." I'd heard many interesting pronunciations over the years but that was a first.

In the beginning the hardest part of boot camp was the sleep deprivation. Getting up early, making the bunk to perfection, making sure uniforms were neat and pressed, and all in a matter of three minutes upon waking. Then we were off to the grinder for physical training, or PT for short. The grinder was nothing more than a really large concrete field that we would all line up on and do more push-ups, sit ups and eight count body builders. I always hated them. An eight count body builder is when a person goes from a standing position and then drops to a squat, immediately followed by kicking their feet out behind them assuming the push-up position. From there they do a push-up and then kick their feet apart and back together again. Lastly they reverse the process ending back in a standing position. As a female with little upper body strength, those were exhausting. Then we would run... and run... and run. It was February and Florida was getting record lows. I remember a girl in our company being from Alaska and even she was complaining about the cold. She told us Florida cold was way different from Alaska cold. Florida cold is a humid cold and the wind cuts through you like a knife getting into your bones. In other words, it was miserable.

We started with 80 female recruits, but little by little the number dropped. For a while, a few of the girls would get anonymous love notes and no one could figure out where they were coming from until one night we woke to a blood curdling scream. A sleepy recruit opened her eyes to find Allyson's unblinking stare inches from her face. Needless to say, Allyson was taken for a psych evaluation and never returned. The love notes stopped after that.

Modesty was something a person needed to get over quickly in the military. Group showers offered zero privacy and the love notes made us suspicious of any glances in our direction while trying to get clean before bed. Once Allyson was gone, our company grew together as a band of sisters. Over the next several weeks, we all got to know each other better and things improved to the point that modesty was no longer an issue.

One of my biggest problems was using the bathroom. I've always needed total privacy to have a bowel movement. There were no doors, no stalls and absolutely no walls between toilets. My solution was to sneak into the bathroom at night just so I could have the room to myself. Constipation was a huge issue for me as well. The stress of boot camp only made it worse. Eventually I started passing blood and then I was forced to get a colonoscopy. In 1989 there was no prepping and no anesthesia. Recruits, males and females, were directed to show up at the Naval Hospital, stick their ass in the air while a male doctor and two male corpsmen stuck the lunar module up their ass. I remember crying because of the pain and humiliation. Nonetheless, it didn't kill me.

We had one member of our company who was tall with a few extra pounds. The company commanders ran that poor girl so hard, but no matter what, she wouldn't quit. Then one morning she fell right out of formation. Couldn't even get up. Turned out she had been running on a stress fracture in her femur and it finally gave way. That was the end of the road for her.

We all looked forward to church every Sunday. Something about boot camp made everyone find God. I think it's because during our most difficult times, we needed something to believe in to keep us going. Even folks who would never darken the doors of a church started attending, if only for an hour of peace where no one was yelling at them. The church was always full, and if you looked around, several of us could be seen sleeping while standing up. I became a Eucharist Minister assisting the priests with Holy Communion. This allowed me to spend additional time at the church. It helped to center me and gave me purpose while finding peace.

Then came work week, the fifth week of boot camp. I can't recall how it happened, but I got assigned the worst job there was, scrubbing the grout on the floors with nothing more than a toothbrush. A few hours after breakfast was served, the line captain, who was in charge of getting the food out to be served, screwed up and was reassigned. In an attempt to improve the situation, I made a suggestion to the third class petty officer in charge. To us lowly recruits, Ricky seemed like a god. He liked my idea and in the blink of an eye, I was done scrubbing grout and officially in charge of the food line. Although a big improvement, I still remember the horrific smell of the giant tubs of powdered eggs. To this day I still can't stand that smell.

Apparently Ricky had taken a liking to me and within a couple days was heavily flirting with me. At one point he followed me into the freezer and attempted to kiss me. I pushed him away and told him how much I needed to get through boot camp and didn't want to risk getting in trouble. I knew he liked me, but I didn't know to what extent. He told me he wanted to take me out once I graduated and out of fear, I agreed, even though I knew my family was coming and I wouldn't follow through.

About halfway through boot camp, I was walking to sick call for a check up and I passed my high school classmate, Charlene. She had arrived not long after I did. She was on crutches and I asked her what happened. She explained that all the running and pounding during PT had taken a toll on her knees. I told her I was sorry and I wished her well. Later I found out she never made it through boot camp, and a few years later she committed suicide. I felt so awful for her and a little guilty because I had made it and she hadn't.

It turned out I was a good marcher. No doubt from years of baton twirling as a kid. They appointed me as a guide on and flag bearer. That person carries a flag and is placed in a position in formation so that everyone to the left and back of them would stay in perfect line. It also allowed the formation to stay in step. I told my parents during one of my few phone calls home and they were really proud of how I was doing.

Phone calls were far and few between, I mostly remember crying because I missed a life of comfort. I hated not sleeping, I was angry because my mom never taught me to sew or iron or properly fold anything. Inspections kicked my butt because I never could get my clothes folded in a way that didn't inspire the company commanders to toss them out of my locker for me to fold again.

About a week before graduation, I received a dozen roses delivered to the barracks. They were from Ricky, the petty officer in charge of the kitchen. He had continued to write me letters telling me about the date he wanted to go on. I continued to ignore the letters for fear the company commanders would figure out what was going on and I'd get in trouble for fraternizing. When those roses arrived, the company commanders, both females, figured out who they were from and called me into their office. They yelled and yelled, interrogating me about having a relationship with a member of the command. I denied everything and swore I had no interest in him. They finally believed me after an hour of interrogating me in the push-up position. When asked why they should believe me, my answer was simple. I can remember my words to this day, "because being in the Navy has always been my dream and I wouldn't do anything to mess that up." They told me there were plenty of other things out in the civilian world I could do, and I responded with a passionate and heartfelt, "NO! There is nothing out there for me and I CANNOT GO HOME! I WON'T GO HOME!" At that point I was hysterically crying in fear that this was the end. All because some horny asshole took a liking to me. Fortunately, at that moment, they believed me. I never heard from Ricky again, and I have no idea what happened to him. I was told that the matter would be addressed because what he had done was wrong.

As the end of boot camp was approaching, my parents were traveling to Florida for my graduation. All of our family that still lived in Florida would be there. Two days before graduation I was scheduled to get the results of my colonoscopy. I told my parents where my appointment was and what time. There they sat in the lobby waiting room when I came in. We were so afraid to hug and embrace because we were breaking the rules. No one was to have any family contact until graduation. With my mom's medical experience, she wanted to be there for the results, so I sat next to her as if I didn't even know her and showed her my medical file. Turns out everything was fine and I got to see my parents two days before graduation. To this day, I wonder, would I have really gotten in trouble for seeing them and hugging them or was I just that dedicated to following the rules and feared getting kicked out? I lean toward the latter.

Finally graduation day came and our company was chosen as the best during pass and review. As a result we got extra leave for the weekend with our families. It was nice to have so much of my family come for the ceremony. I felt bad for the girls who had no one there. Imagine going through what, for many, is the most challenging eight weeks of your life only to finish and see no one you love in the stands supporting you. So often young people join the military not only to serve their country, but to seek something better than what they have with the hope of finding better opportunities. For many it's their only hope for successful future.

Being from Florida, I got permission to travel to the coast with my family and all I cared about was soaking in a hot bathtub. My mom said

I spent the whole day there. It's funny how much you miss the simplest of life's pleasures when you can't have them.

Yeoman A School, Meridian, Mississippi May, 1989

Sunday night I went back to the barracks and waited for my orders to A School. Yeoman School was a five week course in Meridian, Mississippi. Nothing like sending a bunch of young folks to a tiny city in the middle of nowhere after eight weeks of hell. Classrooms all day and drinking all night. Back in 1989, the federal drinking age was 18, which meant if you were on the military base, you could drink before turning 21. I made some great friends and had a crush on this one guy. We dated during our time there, but I wouldn't sleep with him because I knew in a matter of a few weeks, our time would end. Sure enough, Mark ended up in Hawaii and I got orders to Lajes, Azores. We were literally going to opposite ends of the earth.

When they gave me my orders, I was dumbfounded. I had never heard of the Azores. The class instructor walked me over to a world map and said, "See that speck of dust off the coast of Portugal?" to which I replied, "Yes." He said, "That's not dust, that's an island and that's where you're going." He then continued, "But first you have to go to jump school." I'm sure the look on my face was priceless as I exclaimed, "WHAT? WHY??" He continued to explain that the island was too small for a landing strip so everything that went there had to drop in. As the lump in my throat began to form and the tears in my eyes started to collect, he joked and told me he was kidding, but that the island was indeed very small and barely big enough for an airstrip.

I managed to get a quick leave home before I had to fly out to the Azores. There I was met by my friends and family and while my visit was short, it was good. My parents drove me to Charleston, South Carolina to catch my flight out to what I would call home over the next 15 months. During our stay in Charleston, we had an amazing dinner at a restaurant that sat on the water. We took a horse drawn carriage ride through the historic sites. After getting off the carriage and walking a few blocks, I realized I had left my purse on the carriage along with my military orders and military identification. Now understand, when you've just come out of boot camp and A school, this is the equivalent of leaving your first born child abandoned on a city bus. I began to run as fast as I could pushing through crowds to find that carriage. There were so many of them. Finally I found the carriage back at the barn where they were switching out the horses and by some miracle, there was my purse. Nothing like adding a little extra stress to an already stressful adventure.

Mom and Dad took me to the airport the next morning and after lots of tears, I was off. It was June, 1989, I was barely 19 years old and I was leaving my family, my friends and my country for a foreign land where I didn't know a soul. But I was excited! This is exactly what I wanted, to see the world and experience different cultures while serving my country. There would be opportunities to travel to Europe and see places I had only dreamed of. I was ready and boot camp had given me the confidence to do whatever I set my mind to.

Four

❦

Welcome to the Azores

The Azores Islands, Portugal, June, 1989 - August, 1990

After a long flight on a C5 military aircraft, sitting on a cargo net with everything I owned in a seabag, I landed on a tiny island barely big enough to build a runway. The purpose of this military base was to spy on the Soviet submarines during the Cold War. It was a strange arrangement. Here the US Navy was in charge of all the aviation activity, the Air Force managed the base, housing and living conditions while the Army ran the port and sea vessels.

There was one beach, if you could call it that. The island was primarily made of volcanic rock, and the small beach located in the bay had sand that was dredged in to give the locals and military a safe place to go. Otherwise, there was a high risk of falling off rocks or being swept away by waves that were ferociously high when they smashed against them.

All in all it was a beautiful place and unlike anything I or most Americans had ever seen. It was like traveling back in time where the primary source of transportation for locals was a donkey and cart. Few spoke English and the men were extraordinarily protective of their women. As you traveled off the base in the evenings it was common to see young men standing on a sidewalk "courting" a young woman who stood on a balcony or porch with a parent supervising. If a date was permitted, then there was an adult chaperone. It was extremely rare, if ever, that a local woman could associate with an American military man.

The food was excellent and local wine was the drink of choice. The views were spectacular from both the interior mountain tops as well as the coastal areas. For an island that was only 11 by 7 miles wide, the scenery was versatile and breathtaking.

I didn't notice right away that the male to female ratio among the military residents was approximately 20 men to every 1 woman. It wasn't immediately obvious to me because I was assigned as a Yeoman which was a common billet for a female. My entire office staff consisted of five females. A Lieutenant, Chief and three Yeoman Seaman's or YNSN's. We worked directly for the Commanding Officer, or CO and his Executive Officer, the XO. All other personnel worked in two separate buildings, the hangar and the ASWOC, Anti-submarine Warfare Operations Center. The ASWOC could only be entered by those with a top secret clearance due to the operations that were conducted inside. The hangar was where our aircraft were maintained.

The weather in the Azores was unlike any place I had ever been. What many folks don't know is that the islands make up the northern point of the Bermuda Triangle. As a result, winter weather which began in November was horrendous. It was literally like living in a hurricane that lasts six months. It basically had two seasons, wet and dry. It was as if one day God turned on the rain and wind and then six months later he turned it off. During the wet season, which began in November, everyone was exempt from wearing their uniform covers (or hats) because it was impossible to keep them on your head. If you were parking a car, you had to make sure to park into the wind so you could muscle the door open. Otherwise, the wind could take a door right off its hinges and go flying onto the runway.

Everyone knew everyone and their business. Those of us who were single or there without a spouse lived in the same building. It was a co-ed barracks designed much like a college dorm. For every two rooms there was one bathroom shared between them. Next to that or across the hall could be a member of the opposite sex. We all shared a common kitchen, recreational room and TV area. There wasn't much TV watching due to the lack of service. We relied on our family members back home to keep us up to date on what was going on in the world. The base theater ran movies approximately six months to a year after they would show in the states. The internet didn't exist yet, therefore there was no Netflix, video streaming, YouTube or any of the other luxuries we have now.

As a result of a lack of entertainment, everyone pretty much gravitated to the only club on base. Therefore, there was a lot of drinking. If you had friends that lived off base, that was often an option to socialize and party.

Shortly after I arrived, I had no problem being invited out to see and do things. Because it was so small, people were anxious to get to know someone new. During the course of having countless invitations to hang out, I was invited to take a tour of the island by a guy I knew as Keith. Keith was an E6, or Petty Officer 1st Class. He seemed nice at first, confident, funny. He was from New York City I think. Being a southern girl myself, northerners seemed to me not as sweet, more direct which came across as arrogant. He took me for a tour of the island and to some of the most scenic spots. It became clear he was interested in more than just a friendship. I wasn't at all attracted to him because he just wasn't my type romantically. He had thick curly coarse hair and a big bushy mustache. He was pushy and made me feel uncomfortable. I was too much of a southern girl and I liked country boys with laid back personalities. I thanked him for a great day and made sure to avoid any situations that would become awkward as time progressed. I couldn't put my finger on it, but there was just something creepy about him that rubbed me the wrong way.

Over the next month, he began to make snarky remarks to me in passing. I assumed his ego was hurt because I had rejected his many invitations to go out. I didn't think much of it, I was 19 and I figured if I ignored him long enough, he would give up.

A little over a month after arriving on the island, I met with some friends at the club after work one night. I arrived back at the barracks just before midnight. I was alone and had left the club before everyone else because I was tired. I wasn't drunk, but I enjoyed a couple Amaretto Sours, which just made me sleepy after a long week at work.

It was common knowledge that while I had a roommate assigned to my room, she was never there. She had a boyfriend who lived off base and as a result, she spent most nights with him. Occasionally she would come in to get something but I rarely saw her. For the most part, it was nice having a room to myself. No one to argue with or mess up the room when it came time for inspection.

When I walked into the barracks that night and entered the Quarterdeck (or entry area), I noticed Keith was on duty. We all had to take turns standing "The Watch" in various locations to include our barracks. The shift would last about 4 hours and then you would be relieved by the next person on the schedule. The shifts ran from 0200-0600, 0600-1000, 1000-1400, 1400-1800, 1800-2200 and 2200-0200. They scheduled them that way so that if you had the 0200-0600 shift you could still make it to work on time. The watch was responsible for maintaining a log book of who came and went after hours. In addition they held master keys and did patrols every hour checking the doors that were normally unlocked during the day. No one liked standing watch, but we all had to do it.

I recall entering the barracks and as usual, Keith made a smartass remark to me which I ignored as usual. He intimidated me because of his rank. I was only an E2 and as an E6, he was in a leadership position. I had worked so hard to get this far, I wasn't about to mess it up by smarting off to a superior.

I went to my room and went to bed. I had just fallen asleep when I heard someone enter my room. Initially I thought it was my roommate. I thought, "Maybe she had a fight with her boyfriend" or "maybe she needed something." It was pitch dark in my room so all I could see was a faint silhouette as a result of the distant street light coming through my window. The next thing I knew, the person who had entered my room was standing right next to my bed. As I sat up to see who it was, I felt a hand go over my mouth and push me down. I couldn't breathe, scream or open my mouth wide enough to bite. Before I could process what was happening, I felt this person put all their body weight on top of mine, pinning me down. He whispered in my ear to be quiet and just enjoy it. I still remember the fear that left me paralyzed, or at least that is how it felt. I can recall the stench of cigarette smoke which I'd despised the smell of my entire life because of my parents. I immediately recognized his voice and knew it was Keith. I could feel his bushy mustache against my ear as he whispered and told me to stay still and don't fight. As he pivoted his weight to pull down my underwear I realized he already had his pants down and before I knew it, my legs were forced apart. He moved his hand from my mouth to my neck and then I felt him ram himself inside of me. It was terrifying, I don't recall ever being that frightened ever in my life. I can't explain why or how, but suddenly it was as if I wasn't there. My mind just checked out and went somewhere else so I didn't have to acknowledge what was happening to me.

After he was done, he told me not to say a word to anyone. He assured me no one would believe an E2 over an E6 and females who make waves by reporting these sort of allegations only hurt themselves by getting labeled as a troublemaker. Sadly, I knew he was right. It was an unwritten rule as a female that you never complain or cause waves or you would be labeled a troublemaker and shunned.

All I ever wanted to do was join the Navy, serve my country, see the world and do it with pride. I got this far, I couldn't screw that up now. So I decided it never happened. *It was a bad dream,* I told myself. I remember becoming nauseous and vomiting in the bathroom. Then I took a long hot shower, went to bed and when I woke up the next day, it was literally as though I just had a nightmare. Unfortunately my subconscious knew better and this was the beginning of a long, hard road ahead.

In the days and months that followed, I continued to run into Keith from time to time and my stomach would just turn to the point I felt I would vomit like the night it happened. He would look at me with this arrogant smirk as though he had conquered me and there was nothing I could do about it. I got to know his schedule around the workplace, so I altered mine in an effort to avoid him. Without knowing it, I was becoming a different person. I struggled focusing and learning my job. I forgot the most basic responsibilities. I would have emotional outbursts and cry over the littlest things. I often complained of stomach pains and headaches which resulted in frequent visits to the medical clinic. At one point I went to take a pregnancy test because I was late. Stress always had a way of messing with my cycle but at the time I feared he had gotten me pregnant.

Eventually I was referred for a psychological evaluation due to my behavior. There I spoke about the stress of learning a new job. At no time did I report what happened to me, nor was I asked. The psychologist determined I was homesick. I found that to be impossible because I was happy to be away from home and happy to have the opportunity to travel. No one asked me if anything had happened to me and out of fear I didn't tell.

Looking back, I wish I had confided in someone. I wish someone had seen past my brave front and recognized the red flags I was presenting. I wish I had screamed out that night instead of being paralyzed with fear, I wish I would have fought back. I wish there had been training for women and even men so that we would know what to do in these horrific situations so we would have been better prepared if or when they happened. I wish women could have reported sexual abuse without the fear of dismissal, but that just wasn't the case. As it stood, there was an unwritten rule that if you reported any type of sexual abuse, you were labeled and marked as a trouble maker. It would be only a matter of time before you would be forced out of the military and that was the last thing I wanted.

Several months later I was bringing in the mail which was often a very heavy bag. A newly promoted petty officer who was normally a pretty nice guy walked over to me as he saw me struggling with the bag. Shane had been a friend and a genuinely nice guy in the time I had been there, but often when someone gets promoted, a little power goes to their head and that seemed to be the case with him. As I was struggling with the mail bag, I thought he was going to help me, but instead, he walked up behind me, grabbed me by my hips as I was bending over to pick the bag up and thrusted his pelvis against my butt. Without thinking, I swung around to throw a punch. He was able to move away quickly and I missed. I was trembling and fighting tears back with all I had. I felt so violated. Whether he meant it as a joke or not, it clearly struck a nerve. As he walked away laughing, he warned me not to swing at a petty officer who outranked me.

As I walked into the Admin Office trembling and crying, my Chief called me over. She demanded to know what was wrong. I told her I was afraid to tell because I was afraid of getting in trouble. She asked me if I was OK and implored me to tell her what happened and that she couldn't help me if she didn't know what happened. It was at that moment one of the other girls spoke up and said, "What did Shane do? I saw you try to hit him." She had seen a portion of what happened through our office window. I looked at the Chief and told her what he had done and why I reacted the way I did. She asked me how I would like it handled because that would not be tolerated. I told her I just wanted to be left alone and to make sure he knew that actions like that aren't funny and that he should never do it again. She asked me if I was sure and I said yes. Later I was called to the Master Chief's office where I received an apology and was again asked if I wanted further action taken. In my eyes, I knew he was joking around and I knew him to be a good guy. I felt a warning was sufficient. However, at this point I had suppressed the rape so much that I didn't say a word about it because it was my word against his and I continued to be afraid of being labeled a trouble maker.

Eventually I made one good friend. Joanna stepped up to take care of me when my wisdom teeth were cut out. I could tell she was a genuinely caring person with no ulterior motives. We connected effortlessly and always had each other's backs. We remained friends until I transferred. I now regret that I never talked to her again, but at the time, I wanted to forget that Lajes, Azores ever existed.

From the time I decided to join the Navy, I hoped I would get an opportunity to see Italy. Eventually that opportunity presented itself. A flight to Italy was leaving the day before Thanksgiving, and I was planning to be on it. As the plane taxied down the runway, it stopped and then returned to the terminal. It seemed all military was banned from entering Europe for non essential travel due to an increase in terrorist threats. That was it, Italy had just slipped through my fingers. I was heartbroken. As I returned to my office, I ran into the command master chief. He apologized for the circumstances as he knew how excited I was to finally see Italy. It was then he suggested I go home for the holiday. He told me there was a flight leaving in an hour for Charleston, SC. With that said, I decided I would surprise my parents for Thanksgiving. It took me two days to travel home. My flight arrived late at night just after Hurricane Hugo devastated Charleston. I was stranded at the airport along with a couple other service members. There were no more flights out until morning and all the hotels were full. It was cold outside and the only place we could find to lie down was on the luggage carriers. So there the three of us were, sleeping on luggage belts in front of automatic doors. Every time the doors opened the cold November air would blow in and wake us up. It was a long miserable night and all we had were our military coats to keep us warm. The next day I managed to catch a flight to Asheville where I had friends who had agreed to pick me up. They drove me over an hour to my house so I could surprise my parents there. As I walked up the steps to our front door, my mom opened it. To say she was excited would be an enormous understatement. She was yelling and screaming and couldn't even form a word. With all her hysteria, my father got up from his chair to see what all the fuss was about. As he spotted me, he smiled and my mom hugged me so hard I could barely breathe. It was amazing to surprise them like that but I swore I'd never do it again for fear her heart couldn't take it.

COURAGEOUSLY BROKEN ~ 33

Later I took advantage of an opportunity to visit Rota, Spain. A few of us took a long weekend and hitched a ride over. It was a popular destination because we had planes flying there to our sister command on a weekly basis. The XO of the squadron happened to be piloting that day and he offered me the opportunity to sit in the pilot seat. There I got my first (and last) flying lesson. We flew at a low altitude so we could take in the scenery as we entered the Mediterranean. The Rock of Gibraltar was so beautiful. It's a memory and a thrill I've never forgotten. Another thrill on that particular flight was a Portuguese A7 that came up from behind and practically stopped in its tracks off our left wing. Our photographer managed to capture an extremely cool photo of the pilot saluting us along with the flap down at 90 degrees. Later I learned the Portuguese had purchased these aircraft from the U.S. and I suppose the pilot wanted to show us what he could do with it.

Once we landed, we partied and toured the area. There was this tattoo artist that everyone had been talking about. His work looked like a painting popping off of a canvas. I had never had a desire to get a tattoo...until that very moment.

Back home on my last day of high school, a young boy had worked all spring to buy me roses on my last day. They were beautiful. He was just a boy who I later realized had a crush on me. I had been his bus driver my senior year and he was my "little helper." When I received those roses and learned how hard he worked to get the money to pay for them, I was genuinely touched. At the time, it was the nicest thing anyone had ever done for me and from then on, roses held special meaning to me.

So there I was, in Rota, Spain. I visited Hung the artist at the tattoo shop everyday for three days, sober as could be and asking questions while I decided what I wanted. Finally I found a rose that I just loved and that was it, I got my first tattoo. Promise number one broken! Sorry Mom.

On another trip we traveled to France. As we arrived in Lisaent, France, we booked tickets for the midnight express to Paris. We had too much baggage to carry with us to the city, so we rented a small hotel room. No one in our group spoke French and the poor hotel clerk didn't speak any English. She called for a supervisor who didn't speak English either. Before we knew it, there were five hotel personnel trying to understand what we were asking of them. We insisted we only wanted to rent one room. Finally it occurred to us why they were confused. There were six of us and only one of us was a female. This particular group of guys were good guys and I felt comfortable knowing I would be safe with them. The hotel staff was under the impression all six of us wished to share a tiny hotel room with one twin bed. Eventually we were able to explain to them we only needed the room to freshen up and leave our belongings for the night and following day while we traveled to Paris. Everyone got a good laugh once the confusion was sorted out.

We got to our room and everyone showered and got cleaned up. I plugged my curling iron in to fix my hair for the trip, knowing I couldn't on the train or once we got to Paris. No one ever told me the electrical wattage in France was different than in the U.S., so when I put the curling iron to my hair, it literally caught my bangs on fire and when I pulled the iron from my head, my hair went with it. There I stood with my bangs wrapped around the curling iron, on fire and smelling horrific. I screamed and ran to the window, which didn't have a screen, and threw my curling iron and hair right out. As I looked in the mirror, I had a huge bald spot about an inch deep right over my forehead. I cried and cried as the guys laughed hysterically at me. One of them was nice enough to loan me his ball cap which I didn't take off the rest of the trip.

While we killed time waiting for our train, we found a small pub. When the owners discovered we were American Military, they embraced us with such love. They fed us until we couldn't eat another bite. The beer was flowing and it was then that I discovered how much I liked French beer and that it had a much higher alcohol content than anything I had experienced in the States. When it was all said and done, they wouldn't accept our money. They fed us out of love and appreciation. After we staggered to the station to catch the midnight express train, we boarded and found our bunk beds just in time to pass out and wake up in Paris the next morning.

Our navigator mapped out the city and figured out how we could see as much as possible in one day before we caught the midnight express train back to the French Air Base on the coast. We walked all day, climbed the stairs to the very top of the Notre-Dame Cathedral and took in the most amazing scenery from the bell tower and roof. We visited the Eiffel Tower and stopped by the Louvre to say "Hello" to the Mona Lisa. I tried escargot for the first time and loved every minute of the city. We did notice however, the people in the city were not nearly as welcoming as those we met on the coast. Perhaps they weren't as grateful for our help during WWII and had forgotten about what life was like under a Nazi regime? The only negative part of the day was the end, when we stopped for dinner before heading to the train station. I had never been anywhere before you had to pay to use a restroom. As I scrambled to find a franc to put into the bathroom stall door, I laid my purse down which had my camera in it. Once I got the bathroom stall door open, I couldn't reach my purse without the door closing. There was no one else in the room so I took my chance. I heard someone come in and leave. As soon as I got to my purse, I noticed my camera was gone. Fortunately I had my identification and money in my pockets, but all of my photos from that amazing day were gone. I was heartbroken. I bought some postcards of the places we had visited and that was all I had to show that I'd been to Paris.

After returning back to the Azores, it was business as usual. My assignment there was a 15 month tour. When I reached my one year mark I could start calling the detailers and start discussing where I would go next. Detailers are the personnel responsible for reassigning military men and women to their next assignments and making sure the most critical positions are filled with priority. The last person military personnel ever want to piss off is their detailer. Otherwise they would find themselves in Nome, Alaska or Tulee, Greenland.

It was 1990 and the Gulf War was kicking off. I was on duty at the ASWOC the night it all began. When I began my watch at 2000 hours, all was quiet, like any other watch. Then at approximately 2230 hours on a Friday night, in walked the CO, XO, Base General and countless other high ranking officers. There were no windows in the building so I couldn't see what was happening outside, but it was clear, this was not a normal Friday night at the ASWOC.

When my shift ended at midnight I was relieved by the next person on watch who asked me if I had heard what was going on. I said, "No, I've been stuck in here but everybody who is somebody is behind those secured doors." That is when they told me the U.S. was going to War and to go look outside. When I walked outside, I couldn't believe my eyes. Our airstrip which typically had only a handful of aircraft parked, was covered with every type of plane imaginable, all parked so closely to one another that I couldn't help wondering how they would get them all back in the air.

It seems our location was the perfect staging area for those being sent to the Middle East. We had all the refuellers, spy planes, fighters and supplies necessary. It was truly an incredible sight.

In the days that followed we realized the Gulf War wasn't going to last long since our enemies were surrendering quite quickly. We didn't get Saddam Hussein but it seemed as though our country had sent him a pretty clear message not to mess with us again.

That summer an opportunity came up to get SCUBA certified. There wasn't much else to do on that island, so I thought I would take advantage of it. I made it through the classroom courses at night after work and got my pool dives in with our small group on the weekend. Finally we were ready to put what we learned to work. The location of the dive was set off a side of the island where, much like most of the island, was volcanic rock. We would have to carefully climb down the rocks to enter the ocean and it was imperative to do this when the seas were calm. As we swam out away from the rocks, a storm came out of nowhere. Our instructor quickly told everyone to swim back and he lined us up to climb back up out of the water. While waiting my turn, the waves began to push me closer and closer to the rocky ledges and it was difficult to fight against them. As if that was scary enough, something suddenly yanked my regulator from my mouth and I was taking on water. I've always been quite buoyant (thanks to my natural flotation devices God gave me), so I was wearing a considerable amount of weights around my waist which made staying above water difficult. I attempted several times to find either of my regulators, but all I could manage to grab were the tentacles attached to the octopus that had ripped the regulator from my mouth. The next thing I knew, I was being pulled from the water after someone realized I was unresponsive. Needless to say, I never went back to finish the course and it would be 30 years before I would have the courage to try again.

Soon thereafter I finally got an offer for my next duty station. I had been asking to go to Bahrain because at that time, women weren't permitted to go into combat or combat zones. Nonetheless, based on my research, I felt Bahrain was as close as I could get to where the action was without being in a combat zone. I just wanted to be part of the action, wherever that may have been.

Despite my best efforts, my detailer told me there just weren't any positions available in Bahrain. He knew I was looking for excitement and action, so he offered me a position in Panama. It was the summer of 1990 and there had recently been a huge conflict there known as "Operation Just Cause." The U.S. had taken out a dictator who was involved in kidnapping and killing American military and smuggling large amounts of drugs into the U.S. Apparently Manuel Noriega had at one time been a CIA informant but his power had gone to his head and he turned on the U.S., so it was time to take him out and hold him accountable for his crimes.

The detailer told me there was a position with a Naval Special Warfare Unit there. I asked him what that was and he said that I would basically be working closely with SEALs. SEAL, like so many other acronyms used in the military, stands for Sea, Air and Land. I'd never heard of a SEAL before, I had no idea what or who they were, but it was Panama, there was action and it was tropical unlike the place I was, so I accepted the position.

As I hung up the phone and announced I had my orders, my XO was walking by. He was a really good man and very down to earth, especially for a pilot. He asked me where I was going, to which I replied, "a Naval Special Warfare Unit." His mouth dropped open and he exclaimed "SEALs!!! You're going to work with SEALs??!!" I wasn't sure why he was reacting this way, so I confusingly said, "Yes, Sir." He laughed all the way to his office and yelled, "You better start doing push-ups and running now." The XO was known for kidding around so I didn't quite take him as seriously as I should have, because he was right. The true adventure was just about to start.

Photograph of a Portuguese A7 taken from the window of our US Navy P3 Orion aircraft as we flew past the coast of Portugal and toward the Mediterranean Sea at a very low altitude.

Five

❧

Oops! Sorry Mom!

Western North Carolina, August, 1990

After leaving the Azores, I made a quick trip home to visit my parents. The flight to Charleston was long as usual with connecting flights to my small town. It always ended up being a two day trip.

Once finally arriving home in the Blue Ridge Mountains, I spent hours talking to my parents about my experiences. They were far more interested in hearing about them than I was telling them. What I really wanted to say was, "Thank God I got out of that hell hole."

As the years passed, they would make mention of the stories I told them about my time on the island, but I struggled to recall them because along with the rape, I suppressed almost everything else about that place. Friends' names, running with bulls, pissing off the base general during a charity golf game for speaking too loudly, I was unable to recall specific details even though I could vaguely remember something that sounded familiar. It all had become a blur in my mind and I was fine with that. After all, I didn't want to remember.

After spending the day telling Mom and Dad of my adventures, it was finally time for bed and there I stood in my bedroom about to change my clothes. I thought about those promises I made to my mom. The cursing hadn't become a real problem... yet. I definitely hadn't been sleeping around, but there was that little issue of the tattoo. As I stood there thinking about all this I knew it was only a matter of time before Mom would walk in my room while I was changing clothes. No one in my house knocked and it was a given my father wouldn't walk in for fear of an embarrassing moment.

I called my mom into my room and told her I needed to talk to her. I began with preparing her and educating her that times had changed from when she was younger. That women were doing things they had never done before and it was considered acceptable. I told her I had thought long and hard about my decision and that it was planned. As I listened to myself ramble on I thought, *She's going to think I'm pregnant. Well then, a tattoo won't be such a big deal.* I was laughing to myself thinking this was a brilliant plan. Before I could go any further in my lecture, she interrupted me and said, "You got a tattoo, didn't you?" I couldn't believe it! I was shocked! How did she guess this? The next thing I knew she stormed out of my room towards my father when these words came from her mouth, "Dan! You're never going to believe what YOUR DAUGHTER did?!" She proceeded to tell him that I got a tattoo. Then she turned to me and said, "That's OK, laser will take it off." I laughed as I replied, "Mom, I didn't pay $40 for something I wanted to turn around and pay thousands to take it off. I'm keeping it." With that, she stormed out of the room. Once out of earshot, my father smiled, chuckled and said, "Can I see? I always wanted one". We laughed and he more or less let me off the hook.

A year later, while home on leave, mom announced to me she had gotten a tattoo as well. I knew she was full of shit and I called her on it. She insisted she had and she began to tell me she could do anything I could. I called her bluff and said, "what did you get?" She replied, "a mouse." Well that was about the dumbest answer I'd ever heard. She hated small creatures with little feet, especially mice. I continued to call her out on her BS and told her I didn't believe her until she showed me. She began to pull her pants down to the area below her waistband...

Nothing! So she pulled them down further... STILL NOTHING... Finally she pulled them all the way down to her upper thigh and still nothing. Then she said, "Oops! My pussy must have ate it" My parents knew I hated that word and would use it just to annoy me sometimes. Then they would wonder why I was the way I was.

During my surprise visit home over the previous Thanksgiving, I reconnected with a guy I'd always had a crush on during high school. David knew how I felt about him and he would only make it worse by flirting with me, but he would never date me because he was older and knew my father would kill him. This time was different though because I was 20 and an adult. Regardless, we tried to keep our relationship quiet because while I could leave for my next adventure, he had to stay in that small town and deal with any consequences from my crazy father. David actually treated me well. Whether he was faithful while I was away I'll never know. He made it clear that he had feelings for me but knew I needed to pursue my dreams. He told me he knew there was a possibility I would meet someone while in the Navy and if I did, to just be honest enough to tell him and he would understand. At the time, I didn't think that was possible. He treated me better than anyone ever had. I truly grew to love him and we stayed in touch even when I was away. Eventually my parents accepted him and to my knowledge, treated him with the respect he deserved. I never heard any different from him which gave me peace of mind. He understood that my father often made my stays in the home unpleasant and always assured me that I could come to him to get away.

In hindsight I can see why I gravitated toward David. He was a good looking, independent man. He had already been out on his own a number of years. He'd been married, had children and divorced. No one ever knew why he moved back to our small town after divorcing his wife, but I gave him credit for traveling several hours to see his kids whenever he could. He made me feel safe and he always had a calm, cool demeanor, something my father never had. My father had abandoned four boys at a young age when he left his first wife. My mother never even knew they existed until after they were married.

I had been through more than my fair share of heartaches at this point which only pushed me to leave that town even more. It seemed like I had nothing in common with anyone and no one could relate to me and my desire to see the world. Then again, they had no idea what I had been dealing with at home for so many years or that my father was bat shit crazy with paranoid schizophrenia. My mom and I didn't know about his mental illness at the time, only that he was impossible to deal with and made life miserable. Not once though did she ever express a desire to leave him. I would have completely supported her if she had. Hell! I would have taken her to Panama with me. If she had just told me all of the dark secrets going on behind closed doors, but she made a vow for better or for worse, and her Catholic beliefs did not allow for divorce. Unfortunately for her, she chose worse... so she stayed.

It was the month of October and I really wanted to return home for Christmas to spend it with David and my parents. In order to do that, I had to cut my visit short and report to Panama as soon as possible. I called my command in advance asking if that was possible and was led to believe it would be. Unbeknownst to me, the country was still recovering from Operation Just Cause and things weren't quite stable yet. Needless to say, my request didn't exactly make points with my new boss considering most of the folks there were at war the previous Christmas. Had I understood that I would have probably rethought my decision, but it honestly had not occurred to me. I was only 20 years old and in love. Nonetheless, the decision was made. I was off to my next adventure.

Six

⚜

Vale la Pena

U.S. Naval Station, Rodman, Panama 1990-1993

I arrived in Panama and was shown to my temporary housing. I was exhausted and they say when your body is exhausted, it will, at some point, shut down.

Unfortunately for me, mine decided to shut down my first night there. Consequently, I never heard my alarm and slept right through it. I never heard the banging on the door either. What I did hear was a woman's voice softly saying my name as she shook me in bed. It seems my Leading Petty Officer (LPO), an E6, had to get base security to let her in my room to wake me up. Awesome! I was off to making a GREAT first impression. I was embarrassed and humiliated, but Lisa seemed understanding of my situation. We hit it off right away and over the years we have remained friends.

Once I made it to the office to meet the rest of the staff, the Admin Officer, Warrant Officer Aquino, introduced himself. He was a high strung guy, who in hindsight, reeked of insecurity. Looking back, I can understand why. He worked under two Navy SEALs, one of whom, the XO, was a big strong extraordinarily well built man with a very serious professional demeanor. It would be years later, after I got to know him on a personal level, that I learned he was a sweetheart, but back then, we were all scared of him. Lisa and I weren't fans of the Warrant, he played favorites and unless you were there for the invasion, you didn't fit in, you were treated as an outsider. I wasn't surprised when I heard stories of him hiding under his desk crying like a bitch when the shit hit the fan. I guess he was a pencil pusher for a reason.

I suppose that is why Lisa and I bonded, we were both "outsiders." Then there was Christy, a YN3 (or Yeoman Third Class). She was a ball of fire. She was a Korean chick who had been adopted by American parents and raised in southern Louisiana. She had a distinct southern accent and she was well loved by everyone. We all knew she was sharp and going places. She did her best to train me to take her place. I had really big shoes to fill, but my struggle to focus and learn new things continued to be a problem, especially with the Warrant constantly pressing down on me. As Christy trained me to take over her position, the Warrant Officer reminded me daily I was falling short of his expectations and my ability to fill her shoes. Honestly I hated it, he made me feel as though no one would ever be as good as Christy which just added extra unnecessary pressure. It's one thing to push someone to be their best, it's another to compare them to others and tell them they will never be as good. But that was his style and everyone knew how he was.

With that said, having come from where I did, an abusive father and carrying the trauma from the rape, it was hard to stay positive much of the time. The Warrant would write me up for typos on the POD (Plan of the Day) and every other minor thing he could find to criticize. He made me doubt myself on a daily basis.

Lisa and Christy would do everything in their power to help me. They couldn't understand why I had trouble learning new things and grasping basic skills. I was forgetful, moody and extremely emotional.

The rest of the Admin staff more or less did their own thing. Diane was only there for about six months after I arrived and then she left the Navy.

We were all so different from one another with diverse backgrounds, but together we made a good team and got stuff done. The Naval Special Warfare Community is a high speed fast paced environment. The motto was "work hard, play hard" and God knows we did a lot of both.

Outside of our Admin department were mostly operators, or SEALs. From the Commanding Officer all the way down to the youngest guys who were on their first deployment, with their first platoon at their first team. The platoon guys would rotate in and out every six months and often we would see the same faces within a year because they would jump from one platoon to another within the same team. They would start out as fucking new guys or "FNG's" as they were commonly called, but as they gained experience and rank, they would return as leaders guiding the next group of FNG's.

Team Guys were operators, not paper pushers. As a result, they treated those of us who knew how to do paperwork extremely well. They valued our hard work supporting them when it came to their travel claims, getting paid, obtaining a travel visa from an embassy and getting them on a flight in a matter of hours. If it involved paper, they needed us to take care of it and make things happen.. and happen quickly.

Then it came time to "play hard." We busted our asses all week and come the weekend, we liked to party. We would drink in excess every night at the club, but often our guys would have run-ins with the Marines or regular Navy sailors known as "fleet guys." Marines, or "jarheads" as we called them always wanted to pick fights with the SEALs because they had something to prove and wanted bragging rights. In all my time there and in all the bar fights I witnessed, not once did I ever see a SEAL (aka Team Guy or Frogman) start a fight or lose a fight, not once. In an effort to avoid the drama and paperwork that followed an incident such as that, we would often party at the SEAL barracks or at someone's house.

The first party I ever attended was the first weekend I was there. One of the platoon Chiefs was hosting it and Christy and Diane took me. As we pulled up to the home, I looked up at the house, which was built on stilts and I saw a guy whose face was fully engulfed in flames diving from a window better than 30 ft high. Apparently Eric was lighting Everclear on fire as he spit it from his mouth. Unfortunately for him, he was blowing into the wind which caused the liquor to blow back and light his face on fire. It was then the platoon corpsman, Roger, grabbed a frozen steak from the freezer and commenced to beating Eric in the face in an effort to put out the flames. In a moment of drunken stupidity, Eric chose to dive from the second story window thinking that was a better idea. *What in the hell have I gotten myself into?* I thought, *These people are crazy.* Oddly enough, many of us questioned Eric's IQ because of moments like this. He was a soft spoken guy and humble beyond words. Both traits are often misunderstood as weaknesses. As it turned out, Eric later left the Navy and went on to graduate from one of the most prestigious universities in the country. While there, he made the football team as a walk on player. After graduating, he returned to the Navy as a SEAL Officer shortly after September 11th happened. From everything I have heard, he became an outstanding leader. I was and still am incredibly proud of him.

Fortunately for him, Eric was fine after taking the nose dive while on fire. That is often the case with drunks. They can survive traumatic falls and crashes easier than sober people because they are so relaxed on impact. I suppose that was the case here because he stood up, arms in the air and laughed at death.

I met some really cool guys that night. One in particular who would become the closest thing to a brother I've ever had. Alex took a liking to me and we got along extremely well. He warned me of the other guys and not to become a "frog-hog" like so many girls do when they get around Frogmen. He was a big guy, had seen battle in Just Cause and lost friends there. He hadn't had it easy himself over the years and we shared a lot in common when it came to growing up. We just connected and he earned my trust by making me feel safe, something I really needed, even though I didn't consciously realize it.

Unfortunately though, he would be leaving soon since his platoon was nearing the end of it's deployment. He told me not to worry though, he'd be back, he loved Panama and it's women.

After Golf Platoon left, I was invited to a party on Chief's Row. Those houses were occupied by the Special Warfare Combat Craftsman (SWCC) or "Boat Guys." Boat Guys were operators who worked closely with the SEALs. Their job was to insert and extract the SEALs into the combat zones during Operations.

This party was being held for a Chief who was highly thought of. It was December 6, 1990 and it was his birthday. I know the exact date because it's the day I met the love of my life.

The girls and I walked into the house, located on Chief's Row and I wasn't quite sure what the hell to think. First it was guys throwing themselves from two story windows after lighting themselves on fire...now this! One would think I would have gotten used to these antics, but they never failed to surprise me. To my left stood three grown men, drunk as hell, with their pants around their ankles as they danced on a coffee table as though they were the Chippendale's at a ladies nightclub. They looked utterly ridiculous. The shock on my face must have been obvious as I stood there with my mouth gaping open. The next thing I knew, this young kid walked up to me and said, "Don't worry, they are harmless." He then introduced himself as Zach and before I knew it, he and I were in the corner of the dining room talking for hours as though we were the only two people in the room.

Zach looked like he was 12 years old. He was 5'10, dirty blonde hair, a pathetic excuse for a mustache and honestly looked like he was trying too hard to be an adult. I wasn't even remotely attracted to him, but he seemed safe and he was keeping me entertained with his goofy sense of humor.

In the blink of an eye, the party was over and my friends had left me. It was late, after midnight and I was about a mile and a half from my barracks. Where did the time go? Zach told me not to worry, he'd get me home. He ran two streets over to where he lived to get "the rocket." The rocket was a car that the guys had literally built themselves with random spare parts from God only knows what.

A few moments later Zach returned and told me that some of the other guys had already taken the Rocket out... most likely to the Blue Goose, which I later found out was a strip club.

So there we stood, awkwardly connected, not really wanting the night to come to an end because we were having fun, but time stopped for no one. Zach asked me where I was staying and I told him Barracks 71 and pointed to the direction of the other side of our Naval Base. He said that was no big deal and he'd be happy to walk me home to make sure I got there OK. Wow! A real gentleman, *Who knew they actually existed*? I thought to myself.

We talked all the way back and along the walk I encountered bats for the first time. That was the day I learned I was terrified of them. Zach was sweet and suggested we walk on the other side of the street away from the jungle. He assured me they wouldn't hurt us and educated me all about the species of the jungles and what I could expect to see during my tour there. He was incredibly funny and respectful. There was just one problem, he had such a baby face. How could he expect me to take him seriously? When I looked at him I saw the face of a child and that was just so unattractive to me considering I had always gone for older guys. Which then reminded me I HAD A BOYFRIEND BACK HOME!

As we arrived at my room, Zach asked me if he could kiss me goodnight. It was then I told him I appreciated his kindness and company, but I had a boyfriend back home. He said he understood and asked if a kiss on the cheek would be OK. I couldn't say no, he still had to make the walk all the way back to his place and I was genuinely grateful for his kindness. Then he asked if we could hang out some more, to which I agreed. After all, I genuinely enjoyed his company and the way he made me laugh.

Over the course of the next couple weeks, Zach and I hung out together every day. We would spend hours playing gin rummy and spades. I was never in my room except to sleep and he had remained a gentleman. Then one night, just before we were both about to travel home for Christmas, he made his move. By this point I was so confused. Here I had a guy who I loved spending time with but he was so different from anyone I had ever been attracted to. Then he kissed me and sparks flew. He looked at me deep into my eyes and said, "You're gonna fall in love with me, you just don't know it yet." I can't lie, it scared the hell out of me. If he only knew me better. Would he judge me when he learned how screwed up my family was? He was from the Midwest and spoke so highly of his family and how close they were. How would he react to knowing how I really felt about my parents and how I was only going home to appease my mom and see my boyfriend?

We decided to put things on hold while we traveled home and he told me that if I got home and missed him, I would have my answer. I was so confused and I felt so guilty, as though I had cheated on David. Then again, David told me if I found someone I wanted to pursue something with, to just tell him and he would understand. I guess this was what he was talking about. I had some serious decisions to make.

Upon returning home, David picked me up at the airport. I was so happy to see him. It felt so good to be in his arms. He had made arrangements to take me to meet his kids five hours from our hometown. He had never taken anyone to meet his kids, so this was special and spoke volumes. I remember meeting his ex-wife who was incredibly nice to me. His kids were so sweet and respectful. The visit actually went really well. Then that night we returned to our hotel room and he picked up on something being off with me. He continued to ask me if I was OK and I just kept saying "yeah, just a lot to take in." What I was really thinking was *I'm such a schmuck for meeting your kids when my mind is on this other guy who has captured my heart.* I continued to focus on David and tried to forget about Zach but as time went on, it seemed more and more obvious that David and I weren't as compatible as I thought we were. We clearly cared for each other and were attracted to each other, but we had absolutely nothing in common.

I didn't want to break up with him before Christmas Day because I had no idea how he was going to react. A few days before Christmas I told my parents I was in town and had some apologizing to do for not explaining why I hadn't seen them sooner. I knew they were going to give me a hard time for being with David instead of them and I didn't want to deal with the drama, so I kept my presence in town a secret. Unfortunately, small towns being what they are, the secret got out and my mom knew what I was up to. I felt bad for hurting her, but I think she eventually understood my actions. I had a really big decision to make and spending as much time as I could with David was the only way I was going to be able to figure out what I wanted.

David came to the house for Christmas Day and dinner. We took pictures and hung out. It was so uncomfortable and awkward because of my dad, and David knew something was off with me.

A few days later, just before leaving town, David and I had a talk. I asked him if he remembered telling me he would understand if I met someone else. He said he did. He then told me he suspected my strange mood may have had something to do with that. I've always been one to wear my emotions on my sleeve and this was no different. I explained to him that as of that moment, this person and I were no more than friends but that he had expressed an interest in pursuing more with me. I told David I wasn't sure how I felt about Zach until I got to see him again. I proceeded to explain that as much as I cared about him and loved him, I couldn't see a future for us because of our differences and directions. David had no desire to travel the world or go with me on any adventures. He certainly had no interest in being the husband of a woman in the military. With that we hugged goodbye and wished each other well. I was free to pursue whatever it was that Zach and I had discovered together.

Just before New Year's Eve I arrived back in Panama and couldn't wait to see Zach to tell him what I'd done. Oddly enough, he didn't seem to feel the same way. He avoided me like the plague. It was bizarre and it was painful. What had I done? I needed an explanation.

It seemed as though Zach was going out of his way to avoid me. On New Year's Eve as we were all at the club, I became distraught as I had hoped so much to spend it with Zach and he was nowhere to be found. As I walked to my barracks room, I saw him walking out of a room a few doors down from mine. That room belonged to one of the biggest sluts on the base. I literally lost my mind. It was the angriest I had been since I got in that fight with Tammy before I left for the Navy, in fact, my anger for Tammy paled in comparison to the rage I was feeling at that moment. I began to scream and yell, cry and make the biggest scene. What had I done to deserve this? As Zach saw me, he took off running. I went to the whore's door and verbally ripped that bitch to shreds. Then I stormed to my room cussing and screaming all the way. I grabbed two t-shirts I had borrowed from him and began to shred them with a butcher knife. When that wasn't enough, I took them downstairs to the parking lot and began rubbing them through oil puddles left behind by cars. I was out of my head in rage, crying hysterically and honestly, I think I could have killed someone at that moment given the opportunity. I had never experienced rage like that before. At least during that fight in high school I had my wits about me and knew what I was doing, but this? This was straight up bunny boiler shit... and I couldn't control it.

I don't remember much after that, but I was heartbroken. Weeks went by and it seemed like I couldn't avoid Zach. We ran in the same circles and short of locking myself up in my room and never going anywhere with friends, I was kinda screwed. Within a few weeks there was a big party at the officer's club. Everyone knew what was going on between Zach and I and you could cut the tension with a knife between us. Christy took it upon herself to recruit one of Zach's buddies and they cornered us together. Christy gave this big speech about how it was obvious to everyone how we felt about each other and how awful it was to be around us because of what was going on. She instructed us to talk and work it out. At least make peace with one another so we could stop being so miserable to be around. It worked. He swore to me I had jumped to conclusions and that he had not done anything wrong. He apologized for being a jerk after returning from his Christmas leave and said he just had a lot on his mind he was trying to sort out. I told him I had broken up with David because while I was away, I realized I missed him and that it wasn't fair to David to continue things while I wanted to pursue my feelings for him. Needless to say we made up and our relationship became official.

Later that night we went back to his place and I got to experience why sex is commonly referred to as "making love." It was AMAZING!! Then, just as things were getting good, I started to hear the sound of wild animals. I mean, I've heard of the earth moving, but wild animals?? Then, right in the middle of having my mind blown, I looked up and there above the partial wall that separates his room from the living room, in the 12 inch space, I could see the four familiar faces of his roommates. Each and every one of them shitfaced drunk. I lost it. I was fuming mad. Talk about ruining the moment? I jumped up, got dressed while yelling about being disrespected and used for their amusement. Zach did his best to explain he had nothing to do with it and it was the first time I saw him lose his temper. As I stormed out of the house crying, I could hear him cussing them out and it sounded like furniture was being thrown. He was set out to kill every one of them... his brothers.

It took awhile for him to convince me he had nothing to do with their prank and judging by the anger he displayed that night I believed him. I was just so embarrassed and humiliated. Eventually one by one they reached out to apologize for their actions and invited me to the house to make it up to me. They were drunk and stupid and out of line... and they were sorry.

Soon thereafter, my bank screwed me over in a big way. It seemed the bank had a glitch and as a result, my direct deposits were being rejected and returned to the Navy. Back then, with no internet and being overseas, by the time you realized there was a problem, it was too late. When my bank statements finally caught up to me via snail mail, I was severely overdrawn because I had written checks on an account where deposits weren't being made. After all the penalties and late payments were assessed, I was severely in the red and badly broke. No money to eat or anything. Zach and the guys took me in because their house had a kitchen. I agreed to cook and clean in return for food and gas in the car I had just purchased. Zach made my car payments for me and in return, he used my car whenever he wanted. We were together all the time anyway, so it worked out great. It was during that time that I was living with the guys who had become my brothers. The ridiculous shenanigans never stopped. They were always in trouble for partying and antics which made me feel as though I was laughing all the time. A big improvement from my last duty station.

Chief's Row was infamous for the trouble the guys would get into. It was all just fun mischief but most of the civilian folks didn't understand their sense of humor. This would create complaints to the higher ups and as the saying goes, "shit rolls down hill." These guys did everything from turning the street into a slip and slide, courtesy of a fire hydrant, to surfing on the roofs of the cars.

I'll never forget when they took the camels from the Nativity scene and mounted one on the other in a "doggie style position. Considering they were sitting on the main highway just before the base entrance, lots of Panamanians and military personnel got to see them before the Base Captain's wife found out. There was hell to pay for that one.

When they were restricted to the base, they decided to bring the party to them. Since they couldn't have a party without women, they put the word out and before you knew it, there was a line of Panamanian women lined up at the back gate to the base. Each guy would go down and sign the women in one by one. There was no stopping them, where there was a will, there was a way. Never tell a Team Guy or Boat Guy they can't do something, they'll find a way one way or the other.

Once they had made too much "jungle juice" which was a potent punch with enough alcohol to kill the average person. But these guys were professional drinkers and it showed. They all loaded up the truck with the punch and headed across the Bridge of the Americas to a Chief's house. They called his place the "He Man Woman Haters Club," why I'll never know, because they all loved women. I think it was just where they would do their male bonding and no women were allowed, except the Chief's wife. Everyone knew she was the real boss and they loved her. Anyway, once everyone made it home OK, what was left of the punch was poured into the yard, along with the fruit that was in it. The following day there was a flock of dead birds in the yard from eating the fruit.

During softball season, everyone on the base would go to the games. As foolish and crazy as these guys were, they played some mean softball. They were incredibly athletic and I wouldn't have been surprised if they won all their games drunk. After winning the championship game, the guys lined their corpsman's truck with plastic and turned it into a mobile swimming pool. There was never a dull moment with those guys and the memories have lasted a lifetime.

One story I never let Zach hear the end of though, was the time he was driving us home. A friend in the backseat was harassing him about his bad driving. Zach, being the comical smart ass he was, thought it would be a great idea to turn the engine off, pull the keys out of the ignition and hand them to the backseat driver.

Just as Zach said, "Fine, then you drive," I had a flashback to my childhood when my father did something similar in the mountains to conserve gas while gravity pulled us down a mountain. I remembered the terrorized scream my mother let out as my father nearly killed us by almost driving over a cliff. I recalled her pulling up on the emergency break between the front seats as we skid to a stop inches from the edge with no guard rail. I was so little when that happened, but it made an impression on me.

As Zach realized he was unable to steer the car around the sharp curve we were quickly approaching because the steering wheel was locked, we could see the rooftops of the homes below and we were heading straight for them. The curve was sharp and straight ahead was base housing. Someone was about to get an unexpected guest dropping

in, literally. Zach quickly tried to get the key back in the ignition to un-
lock the steering wheel but time was running out. We were all screaming at him which I'm sure didn't help. That was when I remembered
what my mom did when I was a kid. I pulled the emergency break up
and we slid to a stop, inches from going over the edge. I will never forget the look on his face. He looked at me with that boyish look and nervous smile and said, "Oops." He knew I was pissed, in fact I was fuming.
What was worse was that he knew when I was quiet, it was the worst
possible sign of my anger. He was driving and I couldn't kill him. I sat
silently all the way home as he drove VERY CAREFULLY the rest of
the way there, repeatedly saying, "I'm very sorry." While I was clearly
not laughing at the time, it became a story we laughed about for a long
time.

One of the best stories about the guys getting in trouble was
for painting a giant dive flag on the living room wall of their house.
Bright red paint isn't easily covered up and the base housing people
were PISSED. It took them 18 coats of paint to cover that red up and
they got their asses chewed out by a highly respected Lieutenant who
went on to be a SEAL Admiral. He could be the nicest guy in the world,
but screw up and he'd let ya have it. One thing everyone agreed on
though was he was fair, which was why he was so highly respected.

There are countless stories of hilariously funny shenanigans that
no doubt helped heal the traumas of war. Once the guys broke into the
house of another and moved every ounce of his furniture into one bedroom leaving him to figure out how to get it all out. There were giant
slingshots invented that started out innocent enough with water bal-
loons then graduated into flash-bangs in the night.

The worst though was when they put Panamanian fireworks in
a window air conditioning unit of the bedroom where one of the guys
was "having relations" with his girlfriend. I suppose you could say the
earth moved for them that night because Panamanian fireworks might
as well have been C4 when they exploded. Thank God, no one was hurt.

It was always something and someone, no one was safe from the
next prank, or "hazing" as they called it back then. Yes, sometimes it

would get out of control, but there was always a corpsman standing by ready to render aid should things begin to go too far.

I found it funny that our grandparents' generation was known to be the "greatest generation" for how tough they were. It was said back then, that the boats were made of wood and the men were made of steel. In my eyes, these guys were no different. They were strong and they were crazy. Today's generation would have never survived, but what didn't kill them, made them stronger and that was more or less the unspoken philosophy. We were all pushed to our limits, physically and psychologically.

The next several months were amazing. Zach and I were inseparable except when he was deployed. Although we each had our own separate places to live, I could pretty much be found at his place most days. I did his laundry, he shined my boots. I reminded him to return his rental movies, he made me laugh. It was like dating my best friend and it was amazing.

Nonetheless, we still had bumps in the road. We were young, both of us 20. I was only five days older than him as our birthdays were in May. I had trust issues and if he wasn't around, I would have panic attacks that something bad had happened to him and in hindsight, I realize I had become dependent on him because he made me feel safe.

Once he took the car to run some errands and I had fallen asleep after work because I was sick. When I awoke, I was confused about how much time had passed and was adamantly convinced something horrible happened to him. I called everyone looking for him. No one had seen him. Eventually I called the military police and gave them a description of the vehicle. Soon thereafter he came rolling in escorted by base police. It was clear he was very unhappy with me and embarrassed. It seems they accused him of stealing my car and treated him like a criminal. I felt absolutely horrible and he couldn't understand why I was so freaked out over his safety.

Often I would completely overreact to things and it caused tension. He claimed I had trust issues. At the time I couldn't see it but as years would pass I eventually began to understand that he was right.

Once in the course of an argument, I had a panic attack. He got angry and asked me "what the hell is wrong with you?" To which I replied, "I was raped!" It was the first and only time I had told anyone. I told him it happened in the Azores before he met me by a 1st Class Petty Officer and I didn't want to talk about it. I made him swear he would never tell anyone because I was afraid of what people would think. I told him to never bring it up to me again and he didn't.

During one of Zach's deployments I was doing his laundry. I found a letter from a girl back home. It was wrong to read it, but based on the hearts and smell of perfume, I couldn't stop myself. In the letter it talked about how much she missed him and how she couldn't wait to marry him. She spoke of the wedding plans she was making for December and how much she liked the idea of a winter wedding. I lost my mind. I called home crying and my parents told me not to make any rash decisions until I spoke with him. Maybe there was more to the story. So about a week later when he returned, I met him at the dock when his crew returned. I didn't say a word, I knew he was exhausted. He had dropped a tremendous amount of weight in two weeks due to a parasite he got in the jungles.

When we returned to the barracks and he got a hot shower, I waited. I was quieter than usual, it was killing me, but I needed to not lose it because that never got me anywhere. Finally he said to me, "OK, what's eating at you? I can tell." I told him something was bothering me but I didn't want to bombard him so soon. He replied, "let me guess, you found the letter?" I just sat with tears in my eyes waiting for an explanation. That's when he told me, "You remember when we got back from our Christmas break and I was distant?" I replied, "yes." He explained to me that when he got home, everyone was waiting for him and expected him to propose to his high school sweetheart, so he did. He didn't know what direction he and I were going to take when we got back and he didn't expect to fall in love with me. He then told me the letter was a few months old and that he had since broken off the engagement. It made sense and I had no reason to doubt him, so we continued our relationship.

After that, things began to get better. He understood I was fragile and I learned I could trust him with my secrets. Occasionally my in-
securities would get the best of me, but then he would turn around and
do something so sweet and reassure me that he loved me.

I had broken my right wrist thanks to tripping on someone's
feet and was in a cast from my armpit to my thumb. My arm was fixed
at a 90 degree angle. I couldn't drive, write, do my hair for work, pay
bills or anything since I was right hand dominant. Zach took over for
me and even learned how to put my hair up for me everyday for work.
He was just the best and with each day, I grew to love him more and
more.

Most of the time Zach treated me with respect and kindness,
but when it came to romance, he fell short. On Valentine's Day he completely dropped the ball and thought ordering out for pizza was good
enough. Even his roommates gave him shit over that. Then May came
and it was my birthday. We met for lunch and he completely ignored
every hint I dropped that it was my 21st birthday. As the day progressed
I got more and more pissed at him. As we drove home after work you
could cut the tension with a knife. I was fuming. Finally he said, "well
we can go to dinner if you want?" The way he said it, I could tell he was
humoring me which just made it worse, but fine, we needed to eat, so I
guess we would go grab food.

As we got ready to go out, he told me he needed to stop by
Mitch and Michelle's, a couple who had become close friends of ours.
As we stopped by their house, he told me to come inside with him. As
I stood in the living room playing with their dog, out came Zach with
a dozen red roses and Michelle walked out with an enormous birthday cake. Jake and Rose had just arrived and it was then that I realized he had planned this for days. I was completely surprised and so
happy. Then I cried and apologized for being such a bitch all day because I thought he had forgotten. He laughed and said, "that was kinda
my plan." He loved to keep me guessing and that was what made being with him so much fun. We all went to my favorite restaurant, Siete

Martes. To this day it has the best lobster thermidor on the planet. I've never found anything even remotely close to being that delicious.

Then the day I had been dreading arrived, June 5, 1990. Zach had gotten orders to leave Panama. He was being transferred to Virginia Beach, Virginia. I only had a year left in Panama and we both felt confident I could get orders to the Virginia Beach area when my time was up. In the meantime we could write, call and fly. We were at least in the same time zone and flights weren't too expensive. We could make it work for a year until I could get there.

Despite all the love and reassurance, something deep inside me felt it would be the last time I'd ever see him. As we sat in that airport waiting for him to board, I couldn't stop crying to save my life. Zach left me some things to ship home to Nebraska and of course I agreed.

We continued to talk while he was home on leave. He let me know when he received his things and told me how he was going to be in his hometown 4th of July parade. On July 5th, I called him to ask how the parade went. His mom answered the phone and said he wasn't there. When I asked her when he would be home so I could call back, she replied, "Well I'm not quite sure, he's out getting married." I sat there in silence, trying to process what I just heard, so I asked her to repeat herself. She said, "He's getting married so he can take his new bride to Virginia with him." It was in that instant my heart broke into a million pieces. I felt as though I couldn't breathe and I went numb as the tunnel vision set in. After everything we had been through, the secrets shared, the promises made, the perfect memories, laughter shared. Mostly though, the love, the intensely deep love that was so real and deeply mutual. How was this possible? How could I have been such a fool? As my mind raised in the seconds it took for all my thoughts and emotions to process, all I could say when she asked if she could take a message was, "Tell him Donna called and said good luck"

Good Luck?? Good Luck?? I thought to myself as I sat there trying to process what just happened. I was in shock and then the anger set in as though it was a wild freight train out of control.

I called his friends who had become like brothers to me. I yelled and screamed asking them how much fun it must have been laughing at me all these months for being so stupid. Jake was the lucky one to get that phone call and he was not understanding anything I was trying to say. Turned out not a single friend of Zach's knew he had a girlfriend back home or that he was engaged. Every one of them was just as shocked and attempted to reassure me that they thought he and I had a future together and this seemed impossible. Jake went as far as calling Zach's home in Nebraska to make sure I hadn't misunderstood something. He spoke with Zach's father who confirmed the news. After that, Zach's friends became my family. They took me in, got me drunk and let me cry until I couldn't anymore.

For the next month I worked and slept. I didn't go out, I didn't talk, I just cried all the time. It was horrible. A new platoon arrived and low and behold, my brother Alex had returned. He noticed right away something wasn't right. I didn't smile, I didn't talk and I was a ghost after work. The SEAL barracks were immediately across the street from mine. Alex showed up one Friday night beating on my door. I had already gone to bed as I did every night early. I hadn't showered in days, which in Panama was not a good thing as the weather was always hot and humid. Alex informed me I was going out and there was nothing I could do to stop it. So I better get my ass in the shower and get ready, otherwise he's carrying my ass out just as I was and I looked like shit. Alex was 6'3 and weighed 240 lbs of pure muscle. Carrying me out kicking and screaming would not have been a problem for him. In fact, he probably would have enjoyed it. So I did as he said, I showered and put a half-hearted effort into my appearance. Alex took me out with the guys, got me drunk, let me tell my story and when I was done, he lectured me on how "that guy isn't worth my tears." He spent the night telling me all the things I needed to hear to realize I could do better. We both got trashed and when the night was over, he got me home safely. After getting out of the truck and standing in the parking lot talking about how beautiful the stars were, we made a mistake, a drunken kiss. Even though I liked it, I freaked out and pushed him away. Fear overcame me and I literally ran home. Days went by and I couldn't talk to him or look at him. He got pissed at me and when I finally realized what I had done by pushing this great guy away I tried talking to him. I told him I was scared and afraid of ruining our friendship which meant too much to me. He said he understood and he became my guardian angel, always protecting me, even from his own brothers.

Our Unit motto was "*Vale la Pena.*" It would be several years before I understood the true meaning of that motto. At the time I thought it meant "Worth the Fight" or "Worth the Pain." I associated those definitions with all the pain of PT and all the grueling training that was done so that we could have the world's best warriors. While that was true, for me, it would be years before I realized it meant the personal pain and growth that would come. It would mean worth the pain of building the most amazing friendships that would last me a lifetime. Most of all, it meant, worth the pain to become the strong woman I became. I'm grateful for the pain because it prepared me for the challenges ahead.

The time the SBU-26 guys converted the truck to a mobile swimming pool after winning the championship softball tournament.

Seven

❦

Part of the Team

Working with SEAL's is an experience that most people will never understand. Even though I wasn't an operator and worked as part of the support personnel, they always made me feel like I was part of their team. Every person in the unit was considered important and played a vital role in getting the missions accomplished. In order to do that, we did everything together. Every morning we showed up for work and the first thing we did was PT. Each day was something slightly different but there was no escaping calisthenics. When I first arrived there I could do the bare minimum of push-ups. Over the 2.5 years I was there, I became an expert at them, for a female anyway. They say Rome wasn't built in a day, well neither were my biceps. We would do wide push-ups, tricep push-ups, regular push-ups, then of course, the dreaded 8 count bodybuilders... God I hated those with a passion. I'm a big breasted woman and anything that involves bouncing is *no bueno* for me. Pulling both knees up under my chest from a push-up position and jumping up never got easy. Jumping jacks were another killer for me. I could tell instantly who the boob men were and who weren't. Fortunately there was only one who would make comments, but no one was safe from the wisecracks of the old Master Chief. Something about that man could be so offensive, but he got away with it because you knew if the shit hit the fan, he's the one you would want there with ya. I adored him despite the days I wanted to kill him. Once he openly spoke of the breast sizes of the women in the unit. He teased Christy about one day starving her kids because she didn't have anything to latch onto then turned around and teased Lisa about having enough for twins. Then lastly he looked at me and announced no one would starve as long as I was around because I had enough to feed a third world country. I got so angry with him and I told him so. For the next week he made his point by only addressing me in the utmost professional way. No more calling me by my first name, only my title. No more cutting up and joking around, it was business only. After a week, I couldn't take it anymore and I missed the real Master Chief. I went to his office and told him I missed him. It was then I learned that just because someone makes some off colored comments towards someone, it doesn't always mean they don't respect them or dislike them. In fact it is quite the opposite. When someone is comfortable enough to joke around in "colorful ways," it can mean they trust you because if they didn't like you, they wouldn't speak to you at all. He added he would be more cautious when joking around because he meant no disrespect. That was a valuable lesson for me and I appreciated it. Master Chief, whom I later had the privilege of calling by first name, remained my friend for many years and that friendship never wavered.

Something else I learned while I was in that incredibly hot country, I don't sweat. Never have, I just didn't notice it until I got to Panama. Sweating helps your body cool in hot temperatures but when you can't sweat in an environment like Panama, it kinda sucks. After a while the guys noticed this. I would come in from our long runs on Fridays and I'd pass the finish line bone dry and ready to pass out. They started making bets on me with the new guys. They would taunt them, "I bet you Donna can run the course and not break a sweat" to which the new guys would accept the bet and lose.

Some guys were sweethearts. Cheese was one of them. A big guy with a giant heart. Something about him lit up a room when he entered and no matter what kind of mood you were in or how stressed you were, his presence made things better. Literally everyone loved him for his kind heart and genuine personality. Cheese would come to visit me in the office and give me massages when he knew I had over done it or if he had a favor to ask. He was a genuine pro when it came to buttering people up. Even without the massage I would have helped him, but no way was I gonna turn him down when he offered. He was so good at them I'd swear he had magic hands. Cheese was killed several years later in the Gaza Strip while working as a contractor delivering supplies to a children's hospital. I think I cried for a week when I got the news. Everyone loved him because there was just something extraordinary about him. He truly was an angel on earth.

Every morning after the guys would PT, they would shower behind our admin office. The girls had a regular shower we could use indoors, but the guys would share a large outdoor area which had just enough wall to cover their torso area. Lisa and I started taking what we liked to call our "Diet Coke break". We would go out back on the patio and wait for the guys to leave the showers. They would walk from the showers to their platoon huts across our compound and often, they would give us a cheap thrill by "accidentally" dropping their towels. I don't care what some women say, we can be just as bad as men when it comes to seeing a fine specimen of the opposite sex. Most of these guys were built like gladiators and we never got tired of admiring their physiques.

That's the thing about Team Guys, they are beyond confident and comfortable in their own skin. Similar to what boot camp was for me with no privacy, these guys went through hell together for six months in BUDS which stands for Basic Underwater Demolition School. BUDS is the training every SEAL must complete and it is known to be the most challenging and difficult training in the world. Much of their time there is spent naked from the showers that follow getting their asses beat in the ocean surf while washing down all their equipment from the ocean salt. This training happens every day and multiple times a day. Any SEAL will confirm that BUDS makes boot camp look like child's play, because it does.

Panama was hot, really really hot. The last thing any of us wanted to do was wear clothing, but these guys were comfortable enough to let it all hang out whenever they felt like it. Seeing as each of them was a pretty fine specimen of a man, Lisa and I were lucky enough to catch a glimpse of the goods every now and then with no complaints.

Around about the same time Alex came back and helped me get my shit together, the Admin unit got a new Warrant Officer. He was awesome and had spent most of his career in the Special Warfare community. Best of all, he seemed to really know how to get things done right. He didn't need to kiss anyone's ass to look good, he had actually earned his respect the right way.

The Warrant put aside anything the previous admin officer told him and took the time to get to know each of us on his own, our strengths and weaknesses. Then he mentored us so that we could become better. He was a positive motivator and thanks to him, the next time I took the promotional test, I aced it with the highest score in the Navy, resulting in immediate promotion. I'll never forget how he gave me the news.

I was in an adjacent office when I heard him yell my name, unlike he had ever shouted before. As though I had REALLY fucked up. I rushed to his office immediately and stood at attention, worried about whatever it was I screwed up. His speech went something like this, "NEVER! In all my years in the Navy! Have I EVER... EVER... had ANYONE working under me..." Then he left this really long pause, as if he couldn't even say it. I just kept thinking, *Oh My God! WHAT DID I DO?!* Then he said it, "I've never had anyone ace the advancement test! Congratulations Petty Officer." I was dumbfounded as I tried to process what he had just said. Lisa shouted and hugged me. We definitely had reason to celebrate. It seemed I was finally beginning to turn things around.

Partying didn't stop, in fact it picked up. Alex taught me how to make a shot called a B52 Bomber. It was a layer of Kahlua followed by

Bailey's Irish Cream with a tiny bit of Grand Marnier. It was absolutely delicious and I somehow managed to talk the bartender into letting me behind the bar to practice my skills. Unfortunately, I got hammered in the process. The next thing I know, all the guys were leaving to go to the Blue Goose, the strip club we girls were never invited to come to. Honestly, none of us had a desire to go, but it was a great way for the guys to ditch their little sisters when they wanted to go pick up girls.

Unfortunately for me, I wasn't in the mood to go home yet, my party was just getting started. So Curtis, a coworker, offered to give me a ride on the handlebars of his bicycle to the Anchorage Club. That was a popular club on the Army base on the other side of the Canal. We would have to ride across the Bridge of the Americas, an extremely large bridge that connects North America to South America. It seemed like a great idea at the time, while we were shit faced.

As Curtis peddled his ass off heading up that bridge with me barely balancing on the handle bars, cars were honking at us as they went by. Clearly we were making a really big mistake. Fortunately a taxi stopped and convinced us to get in and gave us a ride to the club. When we arrived, we had to enter through the doors that led to a long flight of stairs down to the club. I literally fell and rolled all the way down, somehow landing on my feet. Yet another demonstration of how drunks can get hurt and not feel a thing. The doorman to the club told Curtis I could not enter because I was too drunk. Curtis convinced him I was a stunt woman and that I do that for a living. Crazy as it seems, the doorman believed him and in we went.

As we made it to the bar, I spotted Alex, talking to a Panamanian girl. I got so angry for being left behind (no doubt due to my drunken stupor), that I walked up to him, slapped the hell out of him with tears in my eyes, crying and said, "How could you do this to me? How could you cheat on me while I'm carrying your child?" With that I touched my stomach and acted heartbroken. The look on his face was priceless. His jaw just dropped. At that, the Panamanian chick slapped him and apologized to me and left. I stood there feeling quite proud of myself. He told me he couldn't get angry with me because it had to be the greatest "cock block" of all time. We had a good laugh and he babysat me the rest of the night and made sure I got home OK.

The next morning was a different story. I would get the worst hangovers, but this one was a record breaker. I literally couldn't get out of bed. I had been vomiting into my trashcan all night and all day. Finally my roommate stopped by and I told her to go across the street and find Alex for me and tell him I needed help. My body was beginning to convulse from dehydration and some might say I was in detox, but I'll never know. I just knew I thought I was going to die.

Alex came over and took one look at me and knew I was in trouble. He left to get the platoon corpsman and came back to get me. He scooped me up and carried me to the unit where our corpsman was waiting. They hooked me up to an IV and gave me some O2. Within an hour or so I started to feel better. Just in time to party again that night.

Looking back I realize I definitely was drowning my problems in alcohol. It was so easy to do considering the company I was keeping. I do not blame them, everyone in the military drank and partied a lot. I'm just lucky I was with guys who had my best interest at heart and would protect me from harm. I credit the SEALs for showing me that all men aren't bad, just some. If it hadn't been for the Teams, I probably would have ended up a man hater like many other rape victims.

Eventually we were permitted to go into Panama City to the clubs after things stabilized in the country. This one particular club, Patatues, had a $10 cover charge and all you could drink. I still can't understand how they stayed in business the way military people could drink. We would all go on a Friday night after a hard work week. We would pile in a car and then at the end of the night, whoever was the "least drunk" got to be the designated driver. In hindsight, this was pretty stupid, but we were young and living life in the fast lane. Team Guys live their lives knowing each day could be their last and they live it hard. When you're a part of that group, the philosophy is contagious. I had nothing to lose. No husband, no children, no reason to go home, I didn't care... I just didn't know I didn't care. Thankfully the guys watched over me and made sure I always got home safe.

One crazy night I got the short stick and was nominated to drive home. I wasn't even close to being sober, but I was the least drunk. We went out in a girlfriend's Yugo. A tiny, crappy car that I did not want to drive. Nonetheless, it was my turn. So there we were, three SEALs, my friend, and me. We did fine getting to base, but as I dimmed the lights to enter through the base security gate, I accidentally turned them all the way off as I proceeded on. As a result, I drove across base with no headlights on. Sure enough, base police caught up to me just as I was pulling into the barracks parking lot. I was about to shit my pants in fear. I knew I was going to be in trouble. I had taken my heels off so I could drive and explained that to the officer through the one inch crack in the window. I explained the car wasn't mine and the window wasn't working properly. As he asked me to exit the vehicle, one of the guys in the back seat leaned forward and began to make an ass of himself. That's when one of the other guys obnoxiously said, "She's our designated driver officer! Don't fuck with her." The guys started to act out in a comedic way, causing a ruckus as I dropped my head in fear. I just knew this was it. The officer said to me, "Is that true? You've been putting up with their drunk asses all night just so you could drive them home?" To which I replied, "yes sir." He then backed up and thanked me for making sure they got home safely and left. It was then I told them from now on, we decide who's driving ahead of time. I was never going to drink and drive again. I got lucky, and I knew it.

Once I went out without them with some girlfriends. I had given my military ID to another girl to hold onto because I didn't want to carry a purse and had no pockets. We got separated in the club and a huge fight broke out. I ended up getting thrown over the wall of the club balcony and was holding on to the railing with the ocean below. The tide was out so the only thing below was giant rocks in the darkness, about 20 feet down. Not a good situation. Fortunately, someone reached over the wall, grabbed me by the back and pulled me back up. The club was then evacuated by the police and I ran to a phone to call the guys. They stayed in that night because they had an early morning. Alex and another guy came to get me and smuggled me back on to base because I didn't have my ID. Later I caught up with my friend and told her I'd never go out with her again for disappearing on me.

Alex and the guys were always coaching me how to navigate the city streets. While Panama had stabilized, it still wasn't safe in certain areas. It was common for people to approach your vehicle selling roses or wanting to wash your windshield for money. There were cues you had to look for to know if you were in true danger.

Then sure as hell it happened one night. A friend and I were returning to base and as usual, we had to pass through one of the rough areas. As I sat at a red light, this guy approached. Suddenly I saw him reach for his waistband and I saw the pistol grip poking out. Instinctively, I let out the clutch and hit the gas and sped off, right through the red light. We were seconds from being carjacked but fortunately got away.

Eventually Alex left Panama and returned for the States when his deployment left. I had made a new friend, Nichole, another Yeoman like me from Kentucky. She worked with the Boat Guys and we quickly became good friends. We decided to move in together at her barracks since I couldn't stand my roommate and she didn't want one she didn't like. Nichole and I became like sisters. She was the little sister I never had. Then came Julie, a base Yeoman. We became the three musketeers and did everything together.

We have so many hilarious memories, like the time Nichole let her bulldog mouth overload her chicken shit ass. She bet one of the guys she could beat him in a game of basketball. He played her like a fiddle. The bet was that if she won, us girls got to dress him in drag and take him out on the town as a woman. If he won, he got to shave off her eyebrows. I warned her not to do it, but as usual, she didn't listen. We were all a bit stubborn and had to learn things the hard way. Ron played Nichole so badly. He acted like he had never held a basketball in his hands. He couldn't dribble, shoot or do anything right. Nichole wasn't tall, but she was fierce on the court. The bet was whoever reached 10 baskets first, won. Nichole put 9 baskets on the board and Ron had zero. Then he broke bad on her. He dribbled circles around her and practically jumped over her on his way to the rim. He nailed ten baskets in the blink of an eye. She cried, we laughed as we told her not to take the bet. Ron went easy on her by only plucking stripes into her eyebrows. She was lucky, it was nothing an eyebrow pencil couldn't fix for a while. We still laugh about it to this day.

There were weekends away to the Caribbean side of the country and we would all have the most amazing times together. We had become family. I spent one of the best Christmases in my life with these people because while we picked on each other, laughed at each other, offended each other, at the end of the day, we had each other's backs and we knew we were there for each other. Good days and bad. No backstabbing, no lying and honest to a fault. We always knew where we stood with one another and while we may not have liked what someone said, we respected their honesty because we knew it came with good intentions.

One day I was at the base personnel department. As I stood in line to take care of some checks for the guys, three jar-heads walked in and stood in line behind me. They were wearing "sterile fatigues" which meant their names and ranks weren't on them. They were just plain camo uniforms. As they stood in line behind me, they began to make incredibly crude sexual remarks about me. When I turned around to confront them, they laughed in my face and asked me what I was going to do about it. Clearly I could do nothing, I didn't know who they were and I was alone. I was furious. While I didn't consciously realize I had been in this boat before in the Azores since I had suppressed those memories, it was clear my subconscious did and my anger was coming out in tears.

By the time I got back to my office, I walked in throwing things around my desk, cursing and crying. The Warrant heard me and called me into his office. As I cried uncontrollably, I told him what happened with the jar-heads. He was furious and said he wasn't going to allow one of his YN's to be treated that way. He told me it was a good thing his neighbor and good friend just happened to be the Master Gunnery Sergeant, which in the Marine Corps, was like a god.

It didn't take long for them to figure out which Marines had just cashed their paychecks based on the logbook and who signed just after me. Within 45 minutes, there were three young marines standing at attention at our quarterdeck waiting to offer me a formal apology for their behavior. I didn't want to face them at first, but my Warrant assured me he would be standing by my side and not to worry. As I stood there and witnessed them trembling in fear as a result of what they had done, I got some gratification. It wasn't until they were done, that I turned around and saw every SEAL available, approximately 12-15 of them, standing behind me looking pissed off with their arms crossed. They definitely put the fear of God in these guys and I felt vindicated, which for me, was a really good thing. I learned that there is good and bad everywhere. I learned that while a horrible thing happened to me in the Azores, it didn't make all men bad because here I had a team of men that would have died to protect me and that made me feel safe, even in unsafe environments.

SEALs are probably the most fun human beings any person could have the pleasure of spending time with. Even the most mundane days can turn into an adventure. Sometimes good, sometimes bad, but in the end, no matter what, humor is always a way to work through whatever it is.

Once on Labor Day weekend, we had been eating and drinking at the base club. There was a very tall Budweiser balloon tethered outside. It was roughly two stories tall. As we left the club, Nichole said to me and a group of Team Guys in a very strong Kentucky accent, "I sure do like that balloon. I sure would like to have it." Now let me explain something about Frogmen. No job is too tough for them and the bigger the risk, the happier they are. Honestly none of us had a good reason to have that balloon. It just seemed like a fun crazy idea to them at that moment. So before I knew it, there they were, letting the air out of it, untethering it and rolling it up. They tried to put it in my itty bitty Geo Tracker. Not only did it fill up my Tracker, but there were about 10 feet of it left hanging out the back of it onto the ground. I remember walking away with keys in hand, telling them I wanted no part of this stupid idea. That it would cause way too much paperwork to get them out of this trouble. That was that. Or so I thought.

The next day I was called into my Warrant's Office. As I entered, he was laughing as though he was having trouble forming the words to his question because he knew what he was about to ask me was utterly ridiculous. He said, "You don't know anything about a stolen Budweiser balloon do you?" In his words, "the look on my face said it all." At that, he said, "Don't say a word, I don't want to know. Go see the Master Chief."

Now Master Chief Murphy was a breed of his own with a handlebar mustache and wild eyebrows to match. At the time he was the most senior SEAL still in the Navy having served in Vietnam. Master Chief explained to me the Naval Criminal Investigative Service (NCIS) was investigating the "Grand Theft of a Balloon" and my name had been brought up as a possible lead in the case. I froze in panic. The Master Chief assured me he knew I was a good kid and while I might know something about it, to please be careful what I said while he tried to get to the bottom of things to clear them up.

I was terrified, but I couldn't show it. I told them the truth and swore that I did not take that balloon. I told them I was certain my roommate hadn't taken it either and explained that the guys I was with were just joking around and didn't have a vehicle big enough to take it either. Then I did it, it was time to get even with the jar-heads. I told them I recalled a big deuce and a half truck passing by shortly after I walked away from the club. I suggested it was possible the Marines had taken it since they had a vehicle big enough to haul it away. I was there for hours getting grilled into giving names to which I just played stupid.

Suddenly they got a call on the whereabouts of the balloon. It came in anonymously with an apology and explanation of being a prank. I was released and that was the end of that.

I'm not sure what NCIS did after that with the case, but our guys were in the clear and so was I. They were proud of me for not diming anyone out. They were my brothers and it was an innocent prank. Not worth anyone losing their career over.

At the time I didn't realize how unstable I really was. I was partying hard while working hard. I thought that was our way of life. I had suppressed the Azores so deeply by this point it was as if I had forgotten I was ever even there. Nonetheless, the damage was done, I just didn't know it. My moods would swing and according to others, I had become a bitch to be around. Drunk Donna was fun, sober Donna was a miserable evil bitch. I discovered this when I overheard the guys talking to my roommates. They had started making plans and coming up with ways to ditch me. That hurt, that really really hurt. I ran home in tears and just cried. I couldn't understand why I had become this miserable person.

Later Nichole and Julie found me and out of genuine concern, talked to me. Julie suggested maybe it was my hormones and suggested I see my OBGYN to figure things out. I had been on the pill for a while at that point and it was decided that it would be best to take a break to see if I stabilized. It didn't work, but since I wasn't in a relationship anymore, it didn't really seem necessary to go back on it, so I left it alone. In the meantime I made an effort to be less of a bitch.

At this point, Bill Clinton had been elected into office. It was 1992 and there were so many military cutbacks we were all feeling the pain. Bases around the world were getting shut down and anyone in a rate that was overmanned was being forced out of the military at the end of their contract. Congress had not yet approved for women to go into combat roles which meant no aircraft carriers or destroyers for me. I was in a rate overmanned with too many females and we weren't permitted on combat ships. So I was given the choice to switch to a rating where women were needed on non-combat vessels or get out of the Navy. My job choices were Boiler Technician or Hull Tech. I am not shy about admitting that I am the most mechanically challenged human being on the planet. I explained to the detailer that if they switched me from a YN to a BT or HT, I would single-handedly sink a ship without intention and I didn't want that responsibility.

By this time, I had earned the respect of my superiors for being a hard worker and knowing how to get things done. The Navy didn't care that I had aced the promotion test and they didn't care that I was only going to be two weeks shy of taking the next promotional exam which would have made me eligible to stay in. I had glowing endorsement letters all the way to the top Admiral in the Naval Special Warfare Community. I had volunteered not to transfer to save the Navy money just so I could stay in. It was a wait and see game, but we were doing our best.

Eight

❦

Welcome to the Jungle

Jungles of the Panama Canal, November 1992

In November 1992, I was wrapping up my second year in Panama. I had typed countless after action reports and heard countless stories from the guys from down range on deployments. I had begun to picture in my mind what it was like on these trips. I saw the money they made since I did the finances and I wondered, while they are special to me, are they really THAT special to make this kind of money?

I started asking my bosses if I could go on an Op with them. Not a real one, I knew that wasn't allowed, but a training operation. Our guys trained as though every operation was real. They trained hard, all the time. After two years I had been expected to PT with them every day. I could now do 200 pushups without stopping and as many sit ups. I was still a lousy runner, but no one is perfect. I had been on helo's with them with my feet dangling out the side as we dropped boats into the water and flew between mountains. Once I went up in the helo with them while they jumped from 13,000 feet and I walked right to the edge to watch them fall. I had no fear. I felt I had earned the opportunity to go into the jungle with them and experience first hand what made them so elite.

Initially my request was not met with enthusiasm. Women who thought they could do anything a man could do were not well thought of. Call them women's libbers or whatever, they were not well liked, by any of us. That wasn't me, I didn't beat my chest saying I could be a SEAL, I just had a genuine curiosity of what they did and I wanted to challenge myself. Once I explained this to them, the bosses discussed it and thought it would be a good experiment for all of us. There was always scuttlebutt over whether a woman could make it through BUDs and if Congress would ever allow women to try. So we all considered this a bit of an experiment to see how it would go.

A training op was coming up and it was a big one. It involved every branch of the military as it was a Joint Training Operation. The decision was made that I would deploy into the jungle with five of our guys pretending to be the bad guys. Our job was to pretend to be terrorists in the jungle looking to take out a Luxury Cruise Liner passing through the Panama Canal.

Initially we staged on an island in Gatun Lake. There was a nice little bohia there where we set up and camped for a couple nights. No one had ever taught me how to hang a jungle hammock and I was left to figure everything out on my own. I used enough 550 cord to hang ten hammocks and the guys would just shake their heads at me. Whenever cord was needed they would just cut it from my area and restring my hammock. There was no bathroom, only a 50 gallon drum that had been pushed into the ground to mimic some sort of barbaric toilet. I never complained once... I knew better. I volunteered for this and no matter what, I wasn't going to complain or quit. I wasn't there to prove something to them, I was there to prove something to myself and somehow, in the two years I was there, the "Never Quit" philosophy had been ingrained into me.

I recall enjoying the peaceful view of the lake and the mountains and telling them that I didn't get the big deal. All we had done was hang out and play cards for two days. They laughed and told me to enjoy it while I could, this was merely a staging area until we got word to move into our position in the jungle. Later that night, about an hour before dark, we got word to move. We loaded our stuff up into the boat and worked our way upstream. I was amazed how this boat could navigate through such a narrow stream. The guys had pulled up the outboard motors and were now paddling us into the stream as far as they could until we hit land. By this point it was completely dark. The thick jungle canopy blocked any light from the sky. It was only 1900 hours and we literally could not see our hands in front of our faces.

All of the guys were armed. I was not. Despite the fact that I had been trained by them with firearms and qualified as an expert according to Navy standards, they would not allow me to carry a loaded gun. I suppose they were concerned how I would react if I got scared and honestly, I couldn't blame them.

I had gotten a crash course prior to leaving the base about the dangers of the jungle environment. There was the fer-de-lance snake which is one of the most poisonous snakes on the planet. The jungle was also home to the bushmaster snake, the largest pit viper in the world with an extremely poisonous venom. I once heard it's strike was so powerful it could fracture a man's femur. The scariest of the jungle wildlife though was a beautiful small frog commonly known as the dart frog. It emitted a toxic oil so powerful that once it entered the bloodstream, death would occur mere seconds later, Otherwise, if it entered the body through the skin, a person would need the antivenom within an hour to survive. It is believed the frog is the most poisonous animal on the planet. One drop the size of a pinhead could kill 10 men. In addition to the deadly wildlife, the jungles were home to countless poisonous plants and trees. One of which was known as the black palm. The black palm was a tree which had poisonous spikes sticking out from it's trunk. While navigating through the jungle, we had to be careful not to reach for one or bump into one as we hiked up steep muddy mountains carrying heavy gear.

As we unloaded the boat and began to hike up this mountain in the pitch dark, it became clear early on that we weren't going to be able to get far until sunrise. It was monsoon season and with every step, we sank to our knees in mud and slid backwards. Even with the night vision goggles the guys had, we needed to wait until morning. Whether this would have been the case with or without me, they wouldn't say, but I suspected I was holding them back.

As everyone prepared to settle down for the night, I had a choice to make. The guys made themselves comfortable as they lay down in the mud. I was terrified, I honestly questioned if I would live to see daylight. So without saying a word, I decided I trusted these men with my life and I knew the safest place for me to be, was between them. Whatever was going to get me, it had to go through them first. So I found the two biggest guys there and I sandwiched myself between them. I didn't sleep a wink that night. The sounds of the jungle were terrifying. The howler monkeys sounded like lions roaring inches from my ear. They aren't the cute little monkeys you see at the zoo. These are vicious creatures who scream at you, shit in their hand and throw it at you. They are vile creatures, or at least I thought so.

Finally, day break came and we were on the move early. We hiked straight up this mountain carrying all our gear. I hated Meals Ready to Eat or MRE's as they were called, so I packed a bag full of canned goods like tuna for protein. Not the smartest move considering the weight, so the first opportunity I had to dump some weight, I did. I cut my bag down to one can of tuna a day to survive on and left the rest on the boat to go back to the operations center. I ate the tuna right from the can and to this day, if I see a can of tuna, I think of the jungle.

For every step we took up this muddy mountain, we slid down two. I was literally up to my knees in mud. When the guys found the coordinates where we were supposed to set up camp, it made the bohia we had come from look like the Ritz Carlton.

So there I was, trying to hang my hammock between two trees on a steep hill. I wasn't tall enough to get it where I wanted so one end of my hammock was merely a foot from the ground while the other was over my head. My hammock hung at a steep angle so all night I would slide down and have to push myself back up. I could hear the guys making snide remarks and laughing to themselves. Fortunately for me, I knew I sucked at "roughing it" and was able to laugh at myself... for the time being anyway.

In the nights that followed, a few of the guys would leave camp and conduct surveillance. The rest of us would be left behind to guard the camp.

The goal of the operation was for the good guys to find us hidden in the jungle before we could strike. All day long we would sit around and play spades, crack jokes and tell stories. Since I was now the new FNG, many of the pranks were on me. For example, one of the guys would say to another, "Hey man, keep an eye out for that other fer-de-lance, remember they run in pairs." Then they would watch me start looking around, watching my steps and becoming completely paranoid of my surroundings.

Then there was the conversation regarding which way the toilet water flowed above and below the equator. Funny enough, I didn't believe them that it flowed in the opposite direction south of us, and there was no internet then or way to look it up, so imagine how stupid I felt later to find out they were telling me the truth on that one.

There was no make-shift toilet in the ground at this location. It was in the woods. So, being the only female on this operation, I would get up at the first sign of daylight so I wouldn't have to go far from camp to pee. Then I would hold it all day to go again at dusk. I had limited my fluid and food intake because I didn't want to produce waste. There was no way in hell I was going to poop in the jungle. NO WAY IN HELL!!

Approximately seven days into the operation I apologized to the guys for the way I smelled. I was so embarrassed. I disgusted myself. They laughed and told me not to worry about it since they couldn't smell past themselves.

My hair was long and in the course of the week, the environment had caused it to mat and resemble some likeness to a swamp creature. One of the guys suggested I go take a bath in the stream just over the hill if it would make me feel better, they didn't seem to mind being dirty and smelling like shit. This was the first I had heard of a stream since I dared not to venture far from the camp. My goal at this point was to merely stay alive.

Nonetheless, a bath in a stream sounded pretty good. I grew up in the mountains of North Carolina. I've skinny dipped in rivers and lakes. *I can do this,* I told myself. So I gathered my shampoo and conditioner and a towel I had managed to keep dry and headed off. When I found the stream, I was incredibly disappointed. I could literally pee more fluid than this so-called stream produced. But, it was fresh clean water and beggars couldn't be choosers. As I stripped down to what God gave me, I noticed that my BDUs could literally stand up on their own. They were so caked in mud that they wouldn't bend or collapse.

There I stood, butt naked straddling a tiny stream trying to get my hair washed and rinsed. All the while, worrying if the guys were trying to catch a peek at me in this predicament. To this day I will never know if they did and as the years have gone by, I don't care. I like to think they kept their word as my brothers. But if they didn't, I hope they at least got a good laugh because I'm quite certain I looked ridiculous with my white ass pointing to the sky as I tried to get my head wet in that ridiculous little trickle of water.

Despite my "refreshing bath," my mood continued to decline. I was miserable and it showed. I was not about to complain or quit, I just wanted it over with. I had learned that these guys earned every penny they made and then some and nothing had really even happened yet. No one was shooting at us. No one was trying to kill us... except maybe Mother Nature, but that didn't seem to phase them. Nope, the environment alone was kicking my ass and I could completely understand now why this was no place for a woman. Man I've pissed off a lot of women libtards over the years with that statement but one thing I've learned, not a single one of them have ever walked in my shoes, so fuck them, I speak from experience.

As my mood declined and the inner bitch in me came out, it was one of our biggest pranksters that began to figure out why. The discussion amongst the group was "why was Donna being such a bitch." Rich piped up and said, "I know why!" To which I replied, "Oh really smart ass... why?" He laughed and said, ``You haven't shit the entire time you've been out here have you?" When I admitted he was right, the guys freaked out. You would think that it was a medical emergency. They acted as if I was going to spontaneously explode because I hadn't pooped in a week. I got lectured about how dangerous it was... yada yada yada. To this day I giggle when I think of how important it is to a man that he goes regularly. As a woman, we don't keep track. We go when we go, no big deal.

That night my knights in shining armor arrived. They had scoped out an old abandoned lighthouse with a toilet out on the canal. They took me from the camp that night and broke into the lighthouse so I could poop. I was so happy I literally skipped down the dock back to the boat singing James Brown's "I feel good.. nana nana nana na." We all had a good laugh and it was then that I got a glimpse of why these guys are so comfortable in their own skins. Why they prefer to be naked whenever possible and don't care who sees them.

The next day we got a visitor at the camp, our Training Lieutenant "Mr P" came for a visit and informed us there had been a mistake and as a result, they would have to extend the Op an extra three days. This was NOT good news. I had mentally prepared myself to be out of there in three days, now it was another six??!!! I quickly did the math and realized that "Aunt Flo" was going to be arriving in about four days. What was I to do? I had no feminine products and if I did, they would be covered in mud and ruined from all the rain.

It killed me to confess this, but a girl's gotta do what a girl's gotta do. This was no time to be modest, no one else was, so why should I? If you're gonna run with the big dogs, ya gotta pee in the tall grass...right? Finally I just blurted it out, "If we are gonna be out here an extra three days one of you assholes needs to get me some damn tampons! There, I said it!" Not surprisingly, the laughter and smart ass comments began and then Rich, who with an absolute straight face said, "Hey guys, this is a problem." They laughed and said it wasn't their problem, it was mine. Rich was quick to point out what happens when a dog goes into heat...every male dog in the neighborhood wants to get at her. Rich continued to ask, "What do you think the howler monkeys are gonna do? She's gonna draw in every one of those bastards in a ten mile radius."

I had a girlfriend who once told me she went to a zoo when it was that time of the month. She said the monkeys could smell it and they would all start to masturbate. That's when I lost it. I pictured every one of those vile demon creatures coming after me and there I would be with no gun or way to protect myself. As expected, the tears started and despite every ounce of my power to hold them back, I couldn't. I withdrew into my hammock and sobbed as quietly as I could hoping no one would notice. Like that was realistic. Finally one of the guys caved to the guilt trip. He blurted out he couldn't stand to see a woman cry and wouldn't be a part of it. Then the Lieutenant let me off the hook and told me there wouldn't be an extension. That everything was status quo on schedule and we would be out on time.

I was then offered to leave with the Lieutenant if I needed to. I assured him I had a few days before my cycle would come and I wanted to stay. I explained I hadn't gone through all I had so far to miss the grand finale. Later I would learn I earned their respect for not quitting despite all the crap they had put me through.

Later that night after our guys were out doing surveillance, it was just me and the Chief at the camp. He was on the north end, I on the south. It was roughly 0200 hours when I felt a strong presence that someone or something was near. I knew it wasn't our guys as they hadn't been gone long and if they were back, they would talk. Nope, this was that sixth sense that I felt. SOMEONE was near and I knew it. I didn't see them, hear them, smell them, feel or taste them. Nope, my five senses didn't sense a thing... but my sixth sense was in overdrive. I had to alert the Chief as quietly as I could but I couldn't move. I was frozen in fear, had no way to get to him. All I could do was shout, but I was afraid to. What if whoever or whatever it was didn't know I was there? I didn't want to advertise it. So I whispered, "Hey Chief." No response. Again, "Hey Chief." Nothing. At this point I knew something was very near and I was terrified, so I screamed, "HEY CHIEF!!" At that he yells back, "WHAT?!" I said, "there is someone out there. Someone is here!" He told me I was crazy but I knew I wasn't. I insisted that someone was in our camp. Finally he got his flashlight to prove me wrong and when he did, he realized I was right. Someone had been in our camp and gone through our things. Then in the distance we could hear the subtle sounds of a twig breaking. Honestly I didn't notice it, but Chief had many years of experience doing this so he knew what he was hearing.

When our guys returned to camp we told them what happened. We knew it was a matter of time before the good guys would attack us so we radioed in for our next movement. At dawn, we packed up our camp and moved to new coordinates, this time on much more level ground closer to the canal. The guys briefed me on what to expect and told me if I was smart, play dead so I wouldn't be kidnapped and interrogated. That sounded like pretty sound advice to me, so I followed it.

I was so excited about the shit that was coming to hit the fan, I was like a little kid waiting for Christmas. THIS WAS IT! The big finale, what the last 10 days of hell were for.

It was about 2300 hours and we were laying low waiting. Trying to act normal like we didn't know what was about to happen. It wasn't easy. Then it started. The sound of gunfire coming from the PBR's on the water. The 50 cal, flashbangs, the smoke grenades and then the good guys all painted up in their war paint. It was the most exciting moment of my life at the time. I think each guy walked up to me, pointed his M16 at me and shot me. Just so I would get the full effect. I remember playing dead with one eye open… I didn't want to miss a thing. It was such a bad-ass attack and beyond cool. I was so happy I didn't quit or I would have missed this amazing experience.

Once we all packed up and got back to the base, took our much needed long hot showers and got a good night's sleep, we returned to the command to debrief the next morning. There, I was included, but remained quiet as the experts discussed the Operation. As everyone took turns giving their input, I discovered who it was that sent my sixth sense into overdrive on the side of that mountain in total darkness. It was one of my favorite pranksters, Chad. He looked at me and said, "Do you know where I was when you yelled out to Chief?", to which I replied, "Oh My God! That was you?!" He laughed his big gregarious laugh and said he was immediately under my hammock and that my scream startled him so badly that he almost jumped up from underneath me which I knew would have likely sent me into cardiac arrest. Once we all got done having a good laugh at all the ways that could have gone wrong and the briefing was almost done, the Captain, or Skipper as we like to call him, looked at me and said, "Donna? Do you have anything you would like to add from your perspective, since this was a first for you and for us?" I knew he was patronizing me, but it was in a charming way. I looked at him and said, "No Skipper, but if a woman ever does go into this line of work, they are gonna have to come up with a way to waterproof pads and tampons." We all had a good laugh and I got lots of accolades for hanging in there despite the guys efforts to make me quit.

Later when asked if I thought women belonged in the SEAL Teams I emphatically stated "NO!" Not because I struggled, but because I got merely a peek into what these guys had to do. I came to the conclusion that while men and women are equal, we are different. That's just how we are designed. Men have brute strength and the ability to survive in environments women just aren't designed for. Women as a rule have better communication skills and we are nurturers. Men are protectors. I can see a man getting killed while trying to help a woman who can't keep up. Or a man dying because a woman couldn't carry him along with all the armor and ammo out of a firefight. Men can pee standing up and women can have babies. WE ARE JUST DIFFERENT and that's OK, because we are also equal.

I am proud of the women who have gone on to fly fighter jets, become tank commanders and excelled in other types of combat roles. However, when it comes to roles that require hand to hand combat and brute strength in war zones, I'm against it. There may be a few select women out there that can prove me wrong and good on them. However, I am adamantly against lowering standards in the military just to appease those who want to see women succeed in areas they weren't designed for. I believe in letting the boys have their boys clubs just like women deserve to have their girl time. People need to get over themselves and their egos and just accept that we are different from one another but deserve equal respect. We need to stop trying to prove that we are something we are not and embrace what we are in order to become the best version of ourselves. Then, if someone disrespects us for our gender or race, or rejects us based solely on that, then we have a valid complaint. However, if we are rejected because of our gender or age and the reason is backed by science, we need to ask why we are trying to accomplish something mother nature didn't design us for?

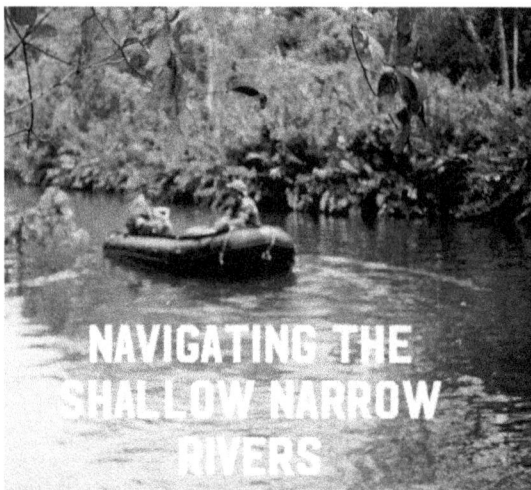

NAVIGATING THE SHALLOW NARROW RIVERS

BLACK PALM - ASTROCARYUM

HOT EXTRACTION

BOTTOM LEFT - JUNGLE PATROLS

BOTTOM RIGHT - TRY HIKING THROUGH THAT

Nine

❧

What are the Odds?

The Unit needed someone to travel to Virginia Beach to take care of some administrative needs and I volunteered to go. I wanted to see it before I left the Navy and I was still hoping to settle there so I could attend Old Dominion University. Alex had allowed me to use his address to apply for instate tuition and offered me a room to rent while I was there. It seemed like a great idea since his other two room-mates were Team Guys and they were always gone. At least this way they would have one roommate who wasn't constantly on the road.

While visiting there for three days, I ran to the base commissary to pick something up and out of the tens of thousands of servicemen stationed in what is the largest Navy base in the world, what are the odds I would run into one specific person? But it happened. As I was going in the doors to the commissary, out walked Zach, literally on the other side of the glass.

I'll never forget the look on his face, he turned white, he looked as though he had seen a ghost. I, on the other hand, was equally shocked. We just stood there staring at each other. Finally he said, "Oh My God! What are you doing here?" To which I replied, "Orders." I couldn't come up with any other words, I was still trying to process the moment.

He said he wished he could talk more but that his ride was waiting and he had to go. He pulled out a small piece of paper, wrote down his office number and told me to call. He asked if I was available for lunch the next day and I hesitated thinking, *What? Are you serious? Why?* But I said yes. I knew why. I needed closure and I still didn't have it.

I immediately called Alex and told him what happened. He coached me how to work through this unexpected invitation and we both agreed meeting Zach to demand an explanation for what he had done might actually give me the closure I desperately needed. Then he told me if the lunch date went any further than that, he would personally kick my ass.

I didn't sleep at all that night, my mind was spinning. Why would he want to have lunch with me after what he did? He's got to know how badly he hurt me. Out of all the people I could run into in this huge place, how is it I ran into him? I mean seriously? What are the odds???

The next day I called him and we met for lunch at a local pub. We talked about Panama, what everyone was up to and about his new assignment. Then I mustered the strength to say what I was there to say, "Zach, I only agreed to this lunch to get an explanation for your actions. How could you? What happened? Why did I have to find out from your mother that you got married?"

He dropped his head, took a deep breath as though he was ashamed of himself and said, "I'm sorry. I didn't know what to do. I felt so pressured and I couldn't figure out how to get out of it." He went on to explain that her parents and his parents were friends and that all of their friends had been together all their lives. Everyone had pressured him to pop the question when he went home on leave that first Christmas shortly after we first met. He said he had broken off the engagement when we got serious, but it seemed she never told anyone, nor did he. So when he got home that summer on leave before going to Virginia, someone came up with the great idea for them to get married at the courthouse so she could travel to Virginia with him. The big church wedding took place that following December as originally planned.

COURAGEOUSLY BROKEN ~ 91

It seemed as though he loved her and had just known her longer. Zach was always a man of his word and I suppose breaking off an engagement via the telephone from another country wasn't exactly the right way to go about things. He didn't want to be "that guy" and he didn't want to disappoint his family by breaking his word. Money had been invested in the wedding and all the plans had been made. All of their friends were counting on them and he just didn't know how to get out of it, or that is what he said.

When I asked him why he couldn't tell me himself and why I had to hear it from his mother, he simply said, "I couldn't bear the thought of hurting you and I didn't want to be responsible for the pain it would cause if I told you." It seemed as though he hoped I would move on and time would heal my wounds, but that never happened.

Alex told me I had my answers and now it was time to move on. No matter what, he never had a nice thing to say about Zach because he was the one to witness my heartache and he was the one trying to help me pick up the pieces. I adored him, but some things are a lot easier said than done. The heart has a mind of its own.

After returning to Panama I began to have recurring nightmares. Every night it was the same one. *I'm living on my own in an apartment as a civilian when suddenly someone comes through a window with a gun and threatens me. I don't know the person, but as he grabs me and lets me know his intentions of raping me, I decide to fight back. I'm not going to go through that again, so I fight. In the course of the struggle, he shoots me in the head. I don't feel any pain, but I feel myself fall to the floor and I hear the blood draining from my head. I feel the blood as it pools under my head and I'm not moving. It's then that I feel myself leave my body and hover over it. I freak out, screaming at the top of my lungs, "Help!" but no one hears me. I run to the street to tell someone what happened to me, but no one sees me or hears me.* Then I would wake up, covered in sweat and crying hysterically. This went on for weeks, to the point that I didn't want to go to bed at night for fear of having the dream again.

Finally I shared this with Lisa and our unit corpsman why I was on edge and having trouble sleeping. They directed me to what I thought was a base psychologist to discuss the matter in hopes I could have the dream analyzed and then once I knew what it represented, it would go away. So I did. Turns out I didn't go to a psychologist but to some sort of substance abuse office where someone there specialized in dream analysis. As a result, this visit would never be entered into my service record. Many years later I discovered A LOT was never entered into my service record.

There the "dream analyzer" told me the dream represented an upcoming change I was concerned about. I told him I was trying to stay in the Navy but there was a chance I wouldn't be able to because of the changes in effect with defense cutbacks. He told me not to worry, it was just stress and would go away. He was wrong, it never did, it continued for many many years.

Soon thereafter, as I was waiting on the outcome of my re-enlistment waiver, I got an answer. On February 2nd, 1993, three days before my contract ended, a decision had been made... DENIED. I was devastated. I thought for sure with the Admiral's endorsement I would get approved, but it seemed the defense budget cutbacks thanks to the Clinton Administration killed that dream.

I left a power of attorney with Nichole to ship all of my belongings home to include my car. Word got out quickly I was leaving soon and everyone was disappointed.

That Friday afternoon, when much of the command was usually gone by 1200 hrs, I was still at work wrapping things up before I left. My presence was requested at the Dive Locker which was unusual. Lisa walked out there with me to see what was going on and as I turned the corner, there stood the entire command, platoons included with a couple kegs of beer ready to send me off. On short notice, the unit put together an amazing plaque, a signed logo from the unit and a ton of food on the grill. I became very emotional as the Skipper made a toast to me and a few others spoke up about how they felt about me and how much my support made their jobs easier. I was fighting tears back, these people had become my family and I didn't want to leave. Then someone said something that was often heard at these types of gatherings followed by everyone echoing the same statement, "SPEECH! SPEECH!" How in the hell was I going to do that when I was barely choking tears back? Then I remembered some advice I once got, make a joke, that always makes things easier.

It was no secret I had no love for the new XO. He was a crotchety old ass who never had a positive word to say about anyone or anything. He would ride my ass (and everyone's) regularly and he loved his coffee strong. If it wasn't like mud, all hell would break loose in the morning as he would yell and shout "I'll do it myself!" I avoided him like the plague, he seemed just like a miserable bastard who wanted to make everyone else miserable. We all gave him a wide berth.

As I began my speech I spoke about how much I loved everyone and how much I would miss them. I told them I would stay in touch and asked that they do the same. As I spoke of this, I went on to say, "I'll miss you all, well, except the XO." At that, I saw jaws drop and then laughter erupt. I saw Lisa pointing over my shoulder and I said, "He's behind me isn't he?" Then everyone was laughing hysterically, including the XO.

After my speech, the XO took me to the side and told me it was nothing personal. He pushes everyone to be their best and being nice was never in his job description. He wished me well and it was genuine. For that brief moment I saw a different, softer side. I saw the human side of him.

Later as I read all the kind words written on my farewell certificate, I read his. It stated, "Donna, Best of Luck in the civilian world and remember, If you can work for me, you can work for anyone. Fair Winds, TC." As the years passed and I went through my fair share of real assholes in the civilian world, I always remembered his words and realized he had done me a favor and prepared me for my future. To this day I have enormous respect for him.

That going away BBQ didn't stop when we left the unit. It continued at the SEAL barracks and everyone was there. We literally partied all weekend and it was amazing. One of the SEAL Chiefs, Tim, played guitar and sang. He had an amazing voice and talent and we all looked forward to him picking up a guitar. We stayed up all night listening to Tim play as the quiet reserved officers who never partied with us got so wasted they couldn't even sit up without falling over. I truly had never felt so loved. It was an amazing send off, one I never have or will forget. I didn't know it then, but as the years went by, the friendships would remain and when times got tough, these people, who had become my family, would get me through my darkest days.

Sunday I flew out of Panama with what few items would fit in my suitcase. It was up to Nichole to get me the rest. I had requested to process out of the Navy in Virginia Beach because I at least knew people there versus Charleston where I didn't know anyone.

I arrived and checked into the Barracks. It was horrible. I felt like I was a criminal in jail. I was surrounded by hateful individuals disgruntled with the Navy and most of them were exiting under "other than honorable" conditions. I was getting an honorable discharge. Why was I in this place? Then I realized I was back in the "regular Navy" and no longer with my brothers and sisters of Naval Special Warfare. I called Alex and explained where I was. He got angry and said I deserved better than that. He had a room that belonged to one of the guys who was deployed and said I could stay at his place while I processed out. There I met his girlfriend and we hit it off nicely. She wasn't the jealous type like so many of the other wives and girlfriends and she understood that Alex and I were nothing more than close friends.

Once I completely processed out on February 5, 1993, my parents chose to drive to Norfolk to pick me up. My dad wanted to visit the bases and reminisce about his days on an Aircraft Carrier. Then we made the long drive home to nowhere, North Carolina, where I felt trapped with none of my belongings or my car.

Shortly after arriving home and a day after getting my car in Charleston, an enormous snow storm labeled "the storm of the century" struck and we lived on that mountain for two weeks without power or water. The roads were too treacherous to drive, so we had to ration what food we had in the house. I was going out of my mind feeling as though there was no escape from the insanity that I had once run from.

It was a toxic environment and all I wanted to do was stay in my room and cry. I had no idea how bad my father's behavior had become and I couldn't fathom how my mother remained there. Every time mom and I would try to talk, he would interrupt and ask us what we were talking about. There were constant allegations of conspiring against him. It was official, he was bat shit crazy. All my mom cared about was not saying something that would set him off.

Having come from an environment where we were encouraged to speak our minds and fix something that seemed broken, I couldn't do it. I couldn't help but call a spade a spade and I had no problem telling my father he was an asshole. It was almost as though he enjoyed the fight and took pleasure in getting one. But it wasn't healthy. These weren't friendly debates. These were vicious insults and tested my temper, which by now I had learned was quite wicked when I let it get the best of me. I remembered calling Alex but found out he was back in Panama again. I called the Unit and begged to speak with him, but he wasn't there. One of the other guys I knew, Gary, took the call and he could hear the desperation in my voice. He listened as I described my dilemma and he recognized I was in a dangerous predicament. Gary told me to just keep my cool and wait until I could get away. He told me to use my time to make a plan to escape and once I did, not to look back. It was just the advice I needed, even though I worried that time wouldn't come fast enough.

Finally the snow melted enough and word got out the highways had been cleared, if we could just get to them. We decided to try and go to the grocery store for food. The only vehicle that could make it out of there was my little Tracker which was 4 wheel drive. We made it to the highway and as we were about half way there, my father's vicious criticism pushed me over the edge. I remember slamming on my brakes in the middle of the highway screaming at him to "get the fuck out!" To which he replied, "What? You can't do that!" I told him it was my car and I could allow or disallow anyone I wished to be in it. I repeated myself, "Get out! Get out! Get out!" I was dead serious. I was hoping a cop would come along to help me, but my Mom, being the peacekeeper she always was, freaked out and begged me to calm down and then told my father he needed to stop. He laughed, he thought the whole thing was funny. But I knew he would throw that temper tantrum in my face soon enough.

I desperately needed to find a job and find a place of my own. My sanity couldn't handle it much longer and I knew I didn't want to live in that town near my father. My mom asked her cousin, who was like an uncle to me, to help me. He had done very well for himself in the Human Resources world and he agreed to help me put together a resume and coach me for job interviews. He lived in Atlanta, approximately three hours from home, so that seemed like a good idea.

Atlanta, Georgia 1993

I went to live with Uncle Kenny and while living with him and helping with his children, he helped me write a resume and coached me on finding a job. It was only a couple weeks until I landed a job working as an administrative assistant at a major Real Estate Law Firm.

I had expressed a desire to buy a gun to my father. He wasn't helpful in a lot of ways, but when it came to firearms, I knew he would support my decision. After seeing how proficient I had become, he was comfortable buying me one for protection. He didn't know what happened to me in the Azores, but he understood and supported my need to feel as though I could protect myself while living alone in a big city like Atlanta. After consulting with my uncle, the retired detective, he bought me a Walther PPK. A small pistol, known for its reliability and compact size. They felt it was the perfect firearm for me. Later I would learn they were wrong. It was extremely difficult to disassemble for cleaning and I just didn't have the hand strength to take it apart. Nonetheless, it worked and would protect me if necessary. I kept it in my nightstand at home and rarely practiced because I didn't want it to get dirty. Not a smart decision, but I didn't want to depend on my father to help me and I wanted to be independent and do it myself. Roughly a year later I traded the gun for one I was much more comfortable with, a Sig Sauer 9mm, just like the one the ole Master Chief taught me on.

In May, 1993 I was signing a lease on my first apartment which I loved. It wasn't far from work and in a great area. Most of the staff at work was really nice except this one girl. She was that nerdy over-achiever no one trusted. She had no problem setting me up for failure and she neglected to explain things to me that would help me become successful. It was clear, she was threatened by me and it showed. She wanted me gone. Nonetheless, the rest of the staff was great, including the attorneys, so I tolerated her in hopes that things would get better.

One day while sending out a bill to a client I caught several charges which were repeated throughout the thick file which caused a significant overcharge. I thought it was a mistake so I brought it to the attention of the attorney. His entire demeanor changed and he said to me that I was never to question their method of billing. In other words, they knew what they were doing and in my mind, that was completely unethical and immoral. So I immediately started looking for another job. It didn't take me long to find one thanks to a great resume and interview skills.

Within a week I found a job at an agricultural company who managed chicken farms throughout the country. There I found myself in a position that had been recently occupied by the woman who had been promoted to the department manager. In essence, I got her old job and she was my new boss. Everyone there was also very nice, but at least twice my age and had been there forever. I just didn't fit in, I shared nothing in common with them.

I wasn't making as much money as I would have liked and I was bored in the evenings, so I took a second job as a cocktail waitress at a comedy club. There I started to have fun while getting paid for it. My coworkers were much closer to my age group and we had a great time. Things were looking up.

When Memorial Day arrived, I went home to visit. My mother managed to get five minutes alone with me to tell me she wanted to leave my father but was worried about how I would feel about it. The only thing that surprised me was that she felt she needed my permission. I looked straight at her and said, "My only question is what took you so long? I've been hoping you would leave him for years!"

I don't remember much from that visit except what my mom told me and I had my first panic attack while shopping in a crowded mall. I couldn't explain it, but I suddenly felt as though I couldn't breath. I began to hyperventilate, felt a sharp pain in my chest and numb tingling sensation in my hands. Tunnel vision set in and I just knew I needed to escape and get some fresh air as quickly as possible. Once I ran out of the nearest exit door, I began to take in several deep breaths and calm down. My mother chased after me asking what was wrong and I just told her I couldn't breathe. Maybe it was too hot or too crowded? Perhaps it was stress? I didn't know, I just knew it scared the hell out of me and I didn't like it.

A few months later while mom and dad were visiting me in Atlanta, it happened again as we were visiting Underground Atlanta. A popular place to shop located below downtown Atlanta. There you could see the oldest buildings in Atlanta which were constructed after the Civil War. Atlanta and it's historic railways had been destroyed by Sherman's troops during the war and the great city began it's comeback at the site where Underground Atlanta lay.

Dad had discovered a jazz and fondue restaurant ironically owned by one of the original Navy SEALs. After dinner at Dante's Down the Hatch, we decided to walk through the underground mall when I felt my heart begin to pound and the tunnel vision closed in. I recognized it as the same sensation that occurred while shopping back home. It became clear I was claustrophobic and when asked why, I blamed it on the octopus that stole my regulator in the Azores. It was crazy enough to believe... Hell! I believed it myself for many years to come.

Ten

When All Hell Broke Loose

Atlanta, Georgia - August 1993

As if separating from the Navy and learning to become a civilian hadn't been hard enough, it was during this time that I began to get a crystal clear picture of how screwed up my father really was.

I spent most of my life just thinking he was an asshole, but as it turned out, he really did have some serious mental health issues. Mom had begun to open up to me, when she could, to tell me the things he had been doing. Accusing her of having an affair with her orthopedic physician. MY MOTHER?? An affair?? NO POSSIBLE WAY!! I knew better. She worked and she coddled him, that was it. She had lost a tremendous amount of weight which he only held against her because he thought she was trying to attract another man. In fact, she just wasn't eating due to stress and was walking everyday in an effort to relieve her stress. Things were deteriorating fast. He had finally healed from his accident years before and was more than physically capable of caring for himself by this point.

Then in August of 1993, the shit finally hit the fan. It was 2am and I was sound asleep when my phone rang. It was my uncle, telling me that my mother had gone missing and my dad was freaking out. My uncle wanted to know if I had heard from her, which I hadn't. He told me my mom had left for the grocery store earlier in the evening and hadn't returned home. He told me to call my dad.

I called my dad to try and figure out what was going on. He was hysterical, rambling on about where she was, she could be dead, what's he going to do without her? Before I could really make any sense of what was going on, my doorbell rang. *Who the hell is at my door at this hour?* I told dad to hold on and he told me not to answer it at this hour. I insisted on answering and he asked me why. I said, "Gee Dad, Mom is missing and I'm three hours away. Don't ya think I should answer it?" He agreed but only if I stayed on the line with him.

As I opened the door, there stood my mom. She looked terrified. All she had was the clothes on her back and her purse. I held my finger up to my lips as though to tell her not to speak, I pointed at the phone so she would understand why. I said, "Hey Dad, Mom is fine, but I gotta go. I'll call you back." When I hung up, he began to call me back continuously and wouldn't stop. I finally turned the phone ringer off so I could concentrate on what my mom was trying to tell me.

She was terrified. Convinced he was going to get in the car and drive to my apartment and kill us both. Neither of us could sleep and I had to work the next day. I got some things together and we left to get a hotel for the night, just so she could relax and we could both get some rest and think with a clear head.

As she began to explain what made her finally leave, she said dad had held her against her will for the past 10 days. He had taken control of the only phone in the house, continued to remain armed and had disabled the car. She said it all came to a head 10 days prior when she woke up with him on top of her holding his .357 magnum to her head. He was convinced she was having an affair and he told her if he couldn't have her, no one could. Then he pulled the trigger, but the gun went *click.* That's when he laughed this sick twisted laugh he had and said, "You're lucky, next time there will be a round in the chamber."

Mom hadn't worked for the past several months due to a number of health issues. Later she would learn many of her health problems were a result of stress over a period of many years. Dad had taken all of their money and put it in an account solely in his name. What he didn't know was that she had managed to acquire her retirement check from the doctor's office where she had worked for years. She managed to talk the doctor into making the check out to just her even though it should have been issued to both her and her spouse according to North Carolina law.

She held onto the check and hid it from dad until she had an opportunity to leave. She showed up at my doorstep with only $11,000 to her name after working 29.5 of their 30 year marriage. Dad got everything, but she got her life.

It quickly became obvious dad wasn't going to give up. I tried talking to him and explaining that she didn't want to come back. He was placing the burden on me to make her, which I wasn't about to do. Finally, after several days of trying to reason with him, it became clear that he couldn't be reasoned with. When I couldn't take it anymore, I raised my voice and yelled at him, "Mom's not coming back and I won't let her. She should have left you a long time ago you crazy bastard!" He lost his mind, started cursing me out and began harassing me non-stop. He was literally filling up two answering machines worth of tape every day. I'd change the tape when I got home and again the next morning before I went to work. Finally I just disconnected it. As it turned out though, those tapes would become valuable evidence.

Not only was dad blowing up my phone at home, he was continuously calling my employer. I was so stressed out and I was getting very little accomplished in what was already a complicated, fast paced job. Finally my boss answered my phone for me and told him I didn't wish to speak to him. She demanded he stop calling the company. He snapped and went off on her. He told her if she didn't let him speak with his daughter, he would blow the company to pieces. That was the first and only time I'd ever been fired from a job. I was devastated.

I called a lawyer to see about a restraining order. I was told they couldn't help me because of the state lines between us. I did however, get some useful advice. The lawyer suggested I go to North Carolina to the county courthouse where dad lived and have him committed to a mental institution. I took down all the notes of what to do and told mom my plan. She said I was wasting my time. "It takes an Act of Congress to get anyone committed," she said. Honestly at this point, what did I have to lose? I was unemployed and in fear for my life and hers. I had to do something.

The next morning I got up really early and made the three hour drive back home. I walked into the sheriff's office with the tapes and a written statement from my mom explaining everything he had done to her.

I filled out some paperwork and within hours, Dad was in custody on his way to the mental hospital a few towns away. It was a temporary solution, but it gave mom and I time to figure out what we were going to do next.

Many people I have come into contact with over the years find themselves in difficult situations because of a toxic family member. Somehow people believe that just because someone is family, they are obligated to put up with the insanity. Life is too short and at some point a person has a choice to make. Either stay in the storm or sail out of it. I chose to sail and leave the drama behind. A parent especially should want the best for their children, not go out of their way to make their life miserable. I never hated my father. I thought I did at one point, but over time, I realized that I felt sorry for him and chose to live my life without him in it if he couldn't be the kind of parent I deserved. He died in 2008 and I felt relief for all of us. My mom could stop worrying about him hunting her down because of the years of threats. I could stop wondering what drama he was going to cause next. He was out of his misery and he went on God's terms at the age of 82 in a state mental hospital.

After Mom left Atlanta, she went to Florida to live with my childhood best friend who was married and had a baby. Dad would never find her there because he didn't know Tina's new last name. Mom then reconnected with several old friends who agreed to help her. She bounced around from house to house until she could get a job and get on her feet. Her brother talked her into getting her real estate license and getting into the timeshare business as a salesperson. He and his family had done really well in the business and she could see that for herself. So that is what she did.

That October, the Atlanta Braves went to the World Series. I had become a big fan of baseball and had gone to several games. Then one night, during the World Series, I was sitting on my couch when my phone rang. When I answered I heard a man's voice say, "Hey Donna, how are you?" The voice sounded very familiar but my mind was playing tricks on me. I sat there afraid to guess, so I just said, "I'm good, how are you?" all the while hoping the person on the other end of the phone would tell me who they were. Then he said, "Do you know who this is?" to which I replied, "I'm not sure, but your voice is very familiar." That's when I heard a name I never thought I'd ever hear again. He said, "This is Zach." I can't recall how long I sat there in silence, unable to form words, trying to process why I was hearing from a guy who had broken my heart, married and moved on. I had just spent the last three years trying to forget about him and now, he's calling me out of nowhere?

Zach went on to tell me how much he missed me and what a huge mistake he had made marrying his wife. He said he thought about me every day and had wondered how I was. When I asked him how he got my number, he said, "I ran into Mitch in New Orleans. He said you had visited him and Michelle in San Diego." When Zach pushed Mitch for information on me, Mitch gave him my number. We spent hours on the phone that night. He was on a brand new ship that had just been built in New Orleans and would be leaving soon for a short deployment before they returned to Virginia.

He called every night and when he was at sea, he would write me letters. He was there for me every night when I would cry about the stress my parents were putting me through. He told me he and his wife were all but over. It was a matter of time before they would divorce. He was miserable.

Meanwhile, I remained in Atlanta since I was still making good money at the nightclub. Then, on December 31st, that came to an end as the company had decided not to renew the liquor license for the club and effective January 1st, we were all out of a job. I continued to look for another good job, but was struggling. I learned Dad wouldn't be in the mental hospital forever and decided Atlanta wasn't the place for me. Mom talked me into coming for a visit to Florida to see what she was doing. She thought it would be a great opportunity for me as well. I had nothing to lose, so I made the move by the end of the month.

Central Florida, February 1994

When I told Zach I was moving to Florida by the end of January, he told me he would be in Tampa, a short drive from where I would be only a couple weeks later. I couldn't wait to see him. I had lost a ton of weight and was the thinnest I'd ever been in my life. I was running everyday and found it to be a great way to channel my stress. I went out and bought this gorgeous black dress, sleeveless, plunging neckline, tight bodice and flowing skirt. I felt like a million bucks when I wore it.

I'll never forget that day, standing on the pier of Tampa Bay as his ship pulled in. I could see the guys on the deck running about preparing to dock. I saw one guy with a pair of binoculars looking at me. I laughed as I could see him looking left and right trying to get the other guys to look. Then one did, he took a long hard look at me through the binoculars. I couldn't tell who was who from that distance, they were all dressed the same in their uniforms.

Minutes later every sailor on that ship was pointing and staring at me. As soon as they could disembark, one began to run down the pier toward me and as he got closer, I could see the huge smile on his face. He walked directly up to me, grabbed me and planted a huge kiss on me. It felt like that iconic moment that was on the cover of *Time* magazine of the Sailor kissing the nurse when WWII ended. It was simply perfect.

I was so caught up in the moment I failed to hear the whooping and hollering coming from the ship. When I got my senses back, I looked up and all the guys were cheering for us. It turned out Zach told them all I was waiting for him and they didn't believe him. They told him there was no way in hell he stood a chance with a woman that hot. I never saw myself as hot. In fact I had always struggled with self esteem, but on THAT DAY, I genuinely stood tall and felt like the most beautiful woman in the world. That was how Zach always made me feel and it never got old. He never said one way or the other, but I think he won a few bucks off the guys that day during their moments of doubt. Either way, I didn't care, I was in his arms again and the next three days were nothing short of amazing.

My mom drove to Tampa one day to have lunch with us. She wanted to meet the infamous Zach that had stolen my heart. It was obvious she did not approve of me seeing him while he was married, but I was all she had and she knew I was too strong willed to listen to her, so she bit her tongue.

After Zach left Tampa, he continued to call and write everyday. I continued to believe he would file for divorce when he returned to Virginia after his deployment was over.

As it turned out, his wife was pregnant with their second child and he failed to mention that to me. He was stuck in the "cheaper to keep her" plan. He had put her through college and now she had decided to be a stay at home mom. He would no doubt be financially screwed because of child support and alimony. Nonetheless, he continued to express his love for me and took every opportunity to see me. It was the beginning of a very long love affair based mostly on emotion since we averaged seeing each other once or twice a year for the next several years.

Despite everything that happened between us, the heartbreak, the lies, the betrayal, he remained my best friend. He was there for me when I needed him most and he knew me better than I knew myself.

No matter what I was going through, his words of encouragement kept me going and he believed in me when I didn't believe in myself. I was just lost without him and the fact that he was hundreds of miles away didn't seem to make a difference. It just made the time we did get to see each other mean that much more.

Between navigating the stress my parents had created and the continuing struggle of finding my way in the civilian world, I worked hard and got my real estate license. No easy feet for a 24 year old who knew nothing about real estate, mortgages and sucked in math. I failed that damn test twice and barely passed it the third time. I went to work for a company that would eventually become one of the largest time-share companies in the world. My cousin had married the owner's son and the entire family was a true rags to riches story. The owner was always good to me and never forgot his humble beginnings despite the incredible wealth he obtained.

Initially I really believed in the concept of timeshare or I wouldn't have sold it. I met so many great people over the six years I worked there who owned multiple timeshares around the world and loved it. I was making decent money and was a top producer earning my million dollar ring, however, after a while, the ether seemed to wear off and I no longer felt satisfied. I needed a change, something to excite me. I was getting bored.

In 1996 a group of SEALs came into town for training. As always I would get no warning, just a phone call out of nowhere that went something like this, "get your girlfriends together and let's party." It never got old. My brothers were in town and fun was about to be had.

I invited a few girlfriends over and we met the guys at a club. It didn't take long before the guys realized it wasn't their kind of scene. Somehow other men immediately viewed them as a threat. SEALs carry themselves with such confidence and they ooze masculinity. Not in a fake gym rat kind of way, but it just came naturally and that is what made them so attractive to women. In an effort to avoid another bar fight before the weekend had even begun, they suggested going somewhere less public to party.

My roommate was out of town for a few weeks on business so I had the place to myself. We all went back to my apartment and things got crazy. There we were, in a two bedroom apartment with the classic Florida retention pond and fountain scenically placed in our view. It's common knowledge that retention ponds in Florida are pretty gross and barely one step above a swamp, but these were Team Guys and they've seen much worse. Before we knew it, one of the quietest guys in the bunch was completely naked running for the pond behind my apartment and making a huge idiot of himself. It was all in an effort to shock my girlfriends and see if they could handle their wild and crazy ways.

I'll never forget what one of my closest girlfriends said, "What the hell is wrong with these guys?" To which I replied, "They know how to have fun! Now you know why I'm always bored when we go out. THIS is my idea of people who know how to party."

We all went into my apartment and drank.. and drank.. and drank. Honestly, I can't remember everything we did other than laugh hysterically at the dumbest shit ever. It was just like old times.

Before I knew it, it was about 4am and I was shitfaced. My mom had gone out of town and her apartment wasn't too far away. I looked around my house and there were bodies everywhere. Seven Frogmen passed out and there was nowhere for me to sleep. One remained, he was the most senior of them all. Scott was a trip, a scary looking dude covered in tattoos but a heart of gold. He and I went to my mom's place and he slept on the couch and I in my mom's bed. He was a perfect gentleman, like the rest of them.

I still remember our conversation that night. I was drunk and talking up a storm about everything my father had put my mother through. Everything he had put me through and how much I hated him. For a brief drunken moment, I asked Scott what it would take to get rid of my dad. He never really answered that question, but instead he posed this to me, "Could you really live with yourself if that were arranged?" In my drunken stupor, I thought the answer was yes, later I valued his response because the good side of me, whatever was left of it, knew he was right.

The next morning, while still somewhat drunk from the night before, one of the guys asked me how far from Deland we were. When I asked him why, he explained there was supposed to be a great drop zone there. Mike had really gotten into sport skydiving. He explained the freedom of freefall without all the weight strapped to him and said it was exhilarating.

I had gone up on a helo with the guys once in Panama and watched them go over the edge. From that moment on, despite my fear of heights, it was something I really wanted to do. I mentioned that to Mike by saying, "Oh My God! I have ALWAYS wanted to do that!" To which he replied, "Well let's go!"

I looked at him as though he was crazy. He explained that they had people there that could take me on a tandem jump. That's when an experienced skydiver straps a first timer to their chest and they go out together in tandem.

We called the place, asked a few questions and the next thing I knew, Mike and I were on our way to the ATM to withdraw $150.00 so I could jump out of a perfectly good airplane.

It was nothing short of AMAZING and the guys were so proud of me. We watched my video over and over that night and they kept telling me most people didn't have the guts to do what I had just done. I was so proud of myself, but mostly I was hooked. I couldn't wait to do it again.

The following weekend I was at the same drop zone taking lessons and in a matter of a couple weeks I was certified as a skydiver.

I jumped every weekend. Every sale I made at work I would take out what I needed for bills and the rest would go to jumps. Each jump was $15 so I would divide everything by 15 and ask myself *How many jumps can I do with this?* or *What do I want more? jumps or dinner?* Jumps always won. I absolutely loved it.

Several months later, Zach came to Florida to do some jump training. The Navy had begun to put Boat Guys through static line jump school for the purpose of launching the boats from planes and heli-copters.

I never thought in a million years Zach and I would be on the same drop zone jumping. The only problem was, he was doing static jumps from a military aircraft and I was doing freefall jumps from a private aircraft. Nonetheless, we would jump all day and make love all night. Life didn't get any better than that I thought.

Several days later, I had just completed my last jump before the sun would set. I was standing there with some friends watching what would have been Zach's last jump of the day. The group exited the tailgate of the aircraft, their parachutes opened and they were falling toward the ground. The spots looked good, the winds were calm, every-thing was fine, until it wasn't. We all stood and watched this one guy as he was rapidly approaching the ground but he wasn't flaring his para-chute. Flaring is something that has to be done just prior to landing. It's kind of like putting on the brakes so that you don't slam into the ground. We were all yelling "FLARE! FLARE! FLARE!" but he didn't. He hit the ground so hard, we literally watched his body bounce. I don't know how I knew it, but this gut wrenching feeling overcame me and I just knew it was Zach. I ran as fast as I could across that airfield and it felt like I was on a treadmill. I just couldn't get there fast enough.

When I finally got there, I ran up to him and my worst fears were imagined. It WAS Zach! He was conscious, but hurt. For the first time in all the years I'd known him and in all that we had been through, I saw fear in his eyes. I choked back tears as he explained he couldn't feel his legs. I just kept telling him I loved him and that he was going to be OK. I remember the thought going through my head, *what if? What if he never walks again?* I didn't give it a second thought, I loved Zach with all my heart and nothing was ever going to change that. All I had ever wanted was to grow old with him and I didn't care under what circum-stances we did that, so long as we were together.

The ambulance came and one of his guys and myself followed it to the hospital. When we got there, they ran several tests and deter-mined he had a significant amount of swelling in his spinal cord. They were admitting him and said we would know a lot more in the following days.

I did everything I could to get the following days off at work, but it was peak season and they just wouldn't let me, so I returned to work the next day. As soon as I could get released from work, I drove to Tampa to see him. I would bring him his favorite food and anything he wanted for the next three days. Finally, he was discharged and returned to his hotel room. Thankfully the paralysis was temporary and although he was in a lot of pain, he was told to take it easy and follow up with his doctors when he returned home. He was supposed to stay in his hotel room and return to Mississippi with the rest of the guys in a few more days.

The next day I drove to the hotel to bring him a home cooked meal since he was getting sick of the same old stuff. When I got there, he was gone. I found one of his guys and they told me he returned home on his own. No phone call, no nothing. I was furious! I called his house in Mississippi and his wife answered. It wouldn't be the first time I had called and she answered. I never told her what my relationship with him was and she never asked. She had to be a fool not to know though. She put him on the phone and I was screaming and crying, accusing him of just using me and not thinking enough of me to at least tell me he was leaving. I had worried myself sick over him and I was hurt. I ended the call with, "I never want to speak to you again."

The Catholic girl in me had struggled with my relationship with Zach, but by this point I was really angry with God. I never stopped believing in Him, I was just really bitter and angry. I suppose in hindsight I had a lot to be angry about, I hadn't exactly had it easy. It seemed the longer our affair went on, the easier it was to justify. It wasn't my fault I fell in love with him before he was married, especially not knowing of his girlfriend back home. If God didn't want me to be with him, then why did he continue to bring him back into my life? I could find a million ways to justify my actions based on my emotions.

Periodically my conscience would get the best of me and I would break things off with Zach in an effort to move on. His shoes seemed impossible to fill, we had too much history and he knew all my secrets, the good, the bad and the ugly. There was no way I would ever find someone I could trust with my deepest secrets the way I had with him.

January 1996 - My first jump

Eleven

❧

Jumping into Love and a New Career

In December of 1996 during the busiest time of year in Florida for skydivers, jumpers would come from all over the world to jump due to the weather and gorgeous skies.

I had become a pretty good relative work diver but not near as good as the pro's who competed across the globe, made movies and jumped for a living. Nonetheless, for a first year weekend jumper, I was getting there. While looking for a group to join, I noticed this extremely handsome British guy who looked a lot like Tom Selleck, cool mustache and all. He walked up to me and asked me if I'd like to join his group as they needed a fourth jumper. I told him I had about 50 jumps and he said, "no problem, you gotta start somewhere, I'll keep it simple." I was starstruck with him. He was good looking, funny and so confident. I could barely speak for fear of looking stupid. So we did our ground work and then got on the plane. The jump went pretty good and Christian gave me some great tips in controlling my falling speed.

Then he suggested we do some two ways so I could practice. We were well into our last skydive of the day, it was sunset, my favorite jump of the day as the skies were so beautiful. Our break off altitude was supposed to be 5000 ft, but instead of breaking off, Christian grabbed me in freefall and planted an amazing kiss onto my lips. The next thing I knew, I was making out at 120 mph plunging toward earth. Then I heard my altimeter beeping in my ear to alert me of my rapid descent. Christian pushed me away and we both quickly deployed our parachutes just before our Automatic Activation Devices (AAD's) did it for us.

When we landed and my head spun from what just happened, Christian ran over to me, both of us laughing and relishing the moment when the dropzone manager came out screaming and yelling at us for our stupidity. Christian took the brunt of it as he was the more experienced jumper. I had to work the next day, so we made plans to meet up for New Year's Eve. I couldn't wait!

A few days later I drove to Deland to meet up with Christian at the DZ. I couldn't find him anywhere. Someone mentioned there was a party on the lake where one of our fellow skydivers lived. Skydiving is not a cheap sport. Between the equipment, jumps and maintenance, it gets pretty pricey. So most skydivers are either wealthy or live a very simple life so they can afford it. It's truly like a drug. Most, including me, would admit they are addicted to the adrenaline rushes and the challenge to improve in order to advance to the next level or technique.

When I found the party, which was held at one of our wealthier friends' homes, Christian was nowhere to be found. No one had any idea where he was. I was beyond disappointed. I don't remember much else about the party other than it was OK at best. I left just after midnight in the hopes I could get home before most of the drunks hit the road.

Life went on, I worked and I jumped. Then a few months later, who showed back up at the DZ? There he was, smiling ear to ear as if he had found his long lost best friend. I asked him why he never showed up that night and he apologized explaining he had gotten food poisoning and was laid up in his hotel all night. He didn't have my number so he couldn't call. He promised he would make it up to me, so we went out that night.

I had the best time. He was funny, we could talk for hours and laugh just as much. The next day we met up again at the DZ and jumped all weekend. He asked if I would give him a ride to the airport and I agreed. When we got to the airport, his flight had been cancelled and the next one was the following day. He had already packed up all his stuff and now had nowhere to stay. So, being the nice hospitable person I am, I invited him to stay with me since I was off the next day. I was completely head over heels for this guy. Something about a sense of humor is so sexy in a man. The next day we got so caught up having fun, we almost didn't get to the airport in time. Christian had bought a new skydiving helmet and he was really excited about it. Men and their toys, no matter where they are from or how old they get, they never outgrow wanting more toys.

Anyway, as we stood in the ticketing area of the airport which was filled with Brits flying home from Florida, Christian was stressing out that he was late. He got snippy with me and I replied to him in a very loud voice, "It's not my fault you're late, you were busy playing with your helmet!" At that moment, every pair of eyes in the lobby turned and stared at me. I felt extremely awkward and as I looked at Christian, he dropped his head, began to turn red and then pulled his helmet from his bag and showed them all what I was referring to. I was really confused as I had no idea what had just happened. Christian looked up at me and said, "They think you just announced I was late because I was playing with my penis." Who knew helmet meant penis in the UK? Later we got a really good laugh and he educated me on all the slang and what meant what. I was told to never call a fanny pack a fanny. It was a belly bag. Apparently fanny is slang for the female nether region.

Over the next several months, Christian and I talked every day on the phone. The time change made it a bit challenging but with my job and getting off early in the day, we made it work. He would send me gifts and cards and I was crazy about him. May was approaching and it was my birthday. Christian wanted to see me but he couldn't get away from work to fly to the U.S. After much discussion, he purchased me a ticket to fly there for my birthday. I was beyond excited.

During the course of keeping up with a long distance relationship, I started working out harder than ever. I was back in great shape and jumping every weekend. I was approaching my 100th jump which is a huge milestone in the skydiving world. It's tradition to do a naked skydive, kind of a right of passage. I put it off as long as possible, but no one at the drop zone was letting me off the hook. Finally a sweet girl from Norway, who had been spending the winter in Florida, was approaching her 100th jump as well. We made a pact. We agreed we would do the naked skydive so long as we were the only two and no cameras were allowed. We made arrangements with the pilot to give us a second pass over the airport after all the other jumpers exited the plane. That's when Kristin and I quickly stripped down to nothing and tied our clothes to our leg straps. All we were wearing were our shoes so we could land safely.

I will tell you I have never laughed so hard in my life. I don't care how low a person's body fat percentage is or how much muscle they carry, the human body is not designed to be seen naked at 120mph. We looked like human shar peis.

Kristin and I landed in a remote field and arranged for another one of our girlfriends to pick us up in a truck. What we didn't plan for was the little league softball game we would have to fly over under our parachutes on our approach to landing. We didn't hear any yelling or laughing from the stands, but we laughed and laughed thinking of the view they would have gotten if they had just looked up.

Finally May arrived and I was off to see Christian. I flew in on a red eye flight and landed early in the morning. Christian picked me up at the airport in his Toyota MR2. It was so weird to sit on the opposite side of the car, especially as he sped down the parking garage ramps at what felt like MACH2. I know he was just trying to get a rise out of me and it worked. I loved the excitement. Finally I just closed my eyes and told him to let me know when we got where we were going. He drove me through London so I could see the sites. He hated the city, but he humored me because we wouldn't be doing any site seeing while I was there. He stopped at Westminster Abbey and I was in awe of its beauty. Television and pictures had not done it justice.

I noticed there were no trash cans anywhere. I walked and walked with trash in my hand and finally complained there was nowhere to dispose of my trash. He told me London doesn't do trash cans anymore due to bombs in the past. He told me to just throw it on the ground and there were sanitation people who would pick it up. Ironically, the city was very clean. I'd say the sanitation workers did a pretty good job considering it's acceptable to litter there.

Finally we made it to his place, a one room flat with a Murphy bed, kitchen and bathroom. He told me what he paid for it and I was perplexed. The cost of living there was outrageous. It made absolutely no sense to me. Turns out living in a place with high taxes really takes a toll on the average person's quality of living. I could see why so many Brits loved visiting America so much.

The next day we drove to Headcorn, a small country village south of London. We stayed in a lovely hotel which had an indoor pool. England was still cold and the weather was constantly changing. The day I arrived it was a beautiful spring day. The day we drove to Head-corn it was a bit warmer and quite humid. We got a few jumps in and we competed in a relative work 4 way competition. We won and I got my first medal. I was the least experienced team member and Christian carried me as he had far more experience than I did. He knew how to adapt to make up for where I wasn't quite as fast as him, but we made it work and I was thrilled.

One thing he wasn't able to do was help me land. I was accustomed to large empty airfields to land on. In England the small airfields are shared with farms which were home to countless sheep and cattle. It was frightening to be coming in for a landing only to see countless sheep and a few cows in your way. Just when you thought you were going to crash into them, they would scurry off leaving you with mild chest pain and a lot of laughter.

That night we drove back to his place, got a shower and he took me out for my birthday. In America if you go out for Chinese food, it's usually some cheap fast food or buffet. That was not the case in England. The restaurant we went to was considered fine dining by American standards. White linen tablecloths, a waiter who never leaves your table should you need anything and the food was the most delicious Asian cuisine I have ever experienced in my life. It was truly an amazing birthday dinner. We both had a bit too much wine to drink which only made us sillier than we normally were. I was having the time of my life.

The next day it was time to go home and I was dreading it. It was like a cold winter day and it actually began to snow as we arrived at the airport. I had experienced four seasons in four days. Christian took me by the market so I could stock up on chocolate. Two things I learned the British do way better than Americans, cheese and chocolate. It would literally melt in your mouth and the flavor was so pure. I still miss the flavor to this day.

As I boarded the plane and found my seat, Elton John's song, Daniel, began to play on my headset as we took off. It struck a nerve and I lost it. I literally cried all the way home to the States. The young gay man next to me felt so awful for me, he comforted me all the way home. It was incredibly awkward for me, but my heart was breaking. Much like when I said goodbye to Zach. I really hated goodbyes.

Christian and I remained in close contact over the next five months while he planned his next trip to the U.S. He was coming for "Bridge Day" in West Virginia. It's a big event as it's the only day in the year when it's legal to base jump off the bridge or repel. I had no desire to base jump, but Christian was looking for the next new adrenaline rush.

I took time off and picked him up in Miami where he flew in. We spent the night in a beautiful resort and the next morning we headed for West Virginia. I never went anywhere without my firearm and Christian didn't know this as the subject had never come up. We got stopped by the police in South Carolina and when the Trooper asked if there were any firearms in the vehicle, I said " Yes sir". I thought

Christian was going to have a heart attack. Being from England, that was something only certain police carried, not even the average police officer was armed. In talking to the Trooper and explaining the situation and letting him know I was a veteran, he was really nice and supported my right to carry. He got a good laugh at Christians' reaction to me having a gun and explained his views and support for the 2nd Amendment. It was quite an education he got and it went against everything he had ever been told about "gun toting Americans." He was hilarious for the rest of the day joking about his armed girlfriend.

Finally we made it to West Virginia and checked into our hotel. We got up early the next morning and I was astonished at the crowds. There were thousands of people from all over the world that came for this event. The New River Gorge Bridge is the world's second longest single arch bridge and the jumpers leap from a height of 876 feet to the river below. When the day was done and the thousands of people left the area, Christian and I hiked back to the top of the mountain and started the long walk to our rental car. It was then we realized we didn't have the keys for it. He insisted he gave them to me, which I had no recollection of. It was quite the quandary. So we called my roadside service who after several hours of waiting, showed up and made us a new key. Christian didn't have a lot of patience in situations like this, but he was grateful I had a plan to get us out of there.

The next stop was on our way to Fort Bragg, NC. Christian wanted to do some jumps with some of the guys from the Army parachute team, the Golden Knights. It was this next stretch of the trip where things began to go wrong. As we woke up in our hotel room the next morning, Christian looked at me and out of nowhere he said, "I want to date other people when we get back to Florida." Apparently he had someone in mind and I knew her. I was furious. One reason was because she had been a friend of mine, the second was because he didn't have the decency to wait until we got back to Florida. There I sat heart broken and stuck with him. I thought about getting a flight home by myself but we were in the middle of nowhere. The next several days, all we did was fight. He got drunk and mean as a snake the night we were in Fayetteville because he was mad he couldn't jump due to bad weather. It felt as though he was taking his anger out on me and honestly it seemed like he was just scared. He was 35 years old and had never been married although he had been engaged twice.

After Christian returned to England, I found out he sold everything he owned, his home, his car, his belongings, quit his job and took all the money to travel the world for a year. Something about needing to *find himself*. A year later he passed through Florida and we ran into each other. No words were spoken, time had not healed a thing.

As usual, I had gained weight like I always did when my heart got broken. All the weight I had lost, the fit body...GONE! I was miserable, beating myself up for falling for a guy that clearly had commitment issues. Why did I always do this?

Time and time again over the next several years, I would start something with a guy who lived in another state or a couple hours away. It was as though I was sabotaging myself, I just didn't know it.

Like so many times before, out of nowhere, the phone rang and it was that familiar voice from the past. It was as if he knew I was in a horrible, vulnerable state. Zach said he missed me and I admitted I missed him too. I told him I couldn't continue to see him romantically though, but I missed his friendship. He agreed to those terms and we started talking regularly again.

By now I was miserable with my job and I was quickly approaching the age of 30. Timeshare wasn't fun anymore and I was tired of stressing over when the next paycheck would come. Working for straight commission is insane and I could never understand how people did it for so many years. I missed the solid paycheck every two weeks, having health insurance and not worrying about bills getting paid. I realized it was time to stop partying and grow up.

I was working a side job at a jewelry store doing cold calls. It didn't pay much but it helped me make ends meet. While working, I met a customer who was a sergeant with the local law enforcement agency. Kind of a goofy looking fella but really nice and entertaining to talk to. As we got to know each other and he learned of my military background, he suggested I look into law enforcement as a career. I laughed hysterically at him. I told him I hated cops as a rule and that he was the only cool one I had encountered in Florida. I told him how I had been treated on traffic stops, being called names and talked down to on several occasions. I told him, in my opinion, the only reason a person becomes a cop is because they have an inferiority complex or were picked on as kids and now wanted to get even with society. He assured me that was not the case and that the persona I had experienced was unfortunate, that yes, there are some jerks out there, but the vast majority really did the job for the right reasons. I laughed and told him I'd pass.

It was Christmas Eve, 1998, and the timeshare industry was slammed as usual. I walked into work miserable as can be and about two hours into the day, I just stopped. I took a long hard look at the room filled with sales people speaking in so many different languages. I saw families that were there from all over the world just to get a few tickets to theme parks with no interest in buying anything. It was loud, I was tired and I just had lost the passion for the job. I walked into my boss's office who wasn't there that day, her assistant was in charge. I told him I was leaving. He flipped out and told me I couldn't leave, it was entirely too busy for anyone to leave yet. I looked directly at him and said, "I'm leaving and I'm not coming back." I had never walked out on a job in my life. This was not planned, it just happened. I couldn't take it anymore. I had no plan, no job lined up and hadn't even started looking for another job. What was I thinking??

As I drove home, my cell phone rang. It was my boss, Gayle. I will never forget her words, "So you quit huh?" I said, "Yep, it's nothing personal, I just can't do it anymore." She laughed and said she understood. Then she asked me what I was going to do and I told her I had no idea as I started to freak out over what I had just done. I told her, "I guess I'm going to go get a copy of the classifieds." Gayle asked me if I would be interested in event planning for conferences and conventions that came into town. I didn't know a thing about event planning, but as it turned out, her best friend owned an event company and was looking for help.

One simple phone call later and I had a job starting Monday. Not too many bosses would do that for an employee that just walked out, but Gayle wasn't your average boss and I was Blessed to have her for one.

As I worked for the event company and learned the ropes of planning and pitching ideas, I got to reap the benefits that came with it. Tickets to the best shows and venues in town so I would pitch their locations to clients who were planning to come. It was a pretty good gig, but it still wasn't me. I needed to find myself. I needed to have a purpose. THAT was what was missing from my life.

Serving in the Navy gave me purpose. Being part of a team that made the world a safer place and stopping drugs from coming into our country gave me pride. Even though the SEALs did all the work, I was behind the scenes helping them do that and they made me feel as though I was valued and appreciated. I needed that in my life and I was at a loss trying to find it again.

I had remained in contact and developed a friendship with the sergeant I had met several months prior. John and I were talking one day and I asked him, "You are constantly complaining about your job and the people you encounter. After 25 years, why do you still do it?" His response rocked me to my core, "Because it's the greatest job in the world. No where else can you be bored for hours or days only to be thrown into seconds and minutes of pure adrenaline and excitement. All the while serving a purpose to help people when they need it most." He went on to talk about the benefits of retirement, steady paycheck, job security, etc. As I sat and listened to him, I realized, that was exactly what I was missing so much about being in the Navy.

As I continued to ask questions and tell him I was thinking about it, he invited me to come on a ride along with one of his female officers. He suggested I ride with a female so I could get a woman's perspective of the job.

I scheduled my ride along for a Saturday night since it would be the midnight shift I'd be riding with. I was welcomed by the squad after being introduced by the sergeant as a friend. I was introduced to Stacy and we hit it off right away. She was a big animal lover as I was. She had recently divorced and was a single mom. She did not act like a cop at all. She was sweet, kind and funny. What surprised me though, was how her personality seemed to change the instant she stepped out of her patrol car to handle a call or interact with a citizen. She was pure business and showed no emotion, about anything. Perhaps this was what John had been telling me from the very beginning? I suppose it is true what they say, "you can't judge a book by it's cover"

We were working in a predominantly Hispanic area and neither of us spoke Spanish. I was amazed how angry Hispanics got when someone didn't speak their language. I could understand if they were here on vacation, but some of them had lived in the U.S. for more than 20 years. How does someone live in a country that long and not speak the language of the land? On top of that, how can they get angry with those who are from there just because they don't speak a foreign language? It always perplexed me and I felt sorry for the officers who got cursed at so much for merely doing their job while trying to help someone that called them.

Nonetheless, we had a lot of laughs. I remember saying, "I always knew there were stupid people in this world, but I had no idea how many." People would lie straight to our faces with the evidence right in front of them. It was a challenge trying to investigate a crime when all some people wanted to do was lie either for themselves or their friends.

Then there were the spiteful ones who would lie just because they were mad at someone and wanted them to go to jail to get even. Everything from domestic violence to rape to theft. It never ended. Stacy said, "It sucks, but as long as there are people like this, we will always have job security." Truer words were never spoken.

The funniest call of the night though was the cab driver who drove a drunk home. The drunk told the cab driver he would be right back with the money to pay him. The cab driver sat and waited for thirty minutes before calling police for "theft of services." Stacy and I walked up to the door to make contact with the guy. We rang the doorbell, pounded on the door, looked in the windows, everything we could think of. Then we heard it, the sound of VERY loud snoring. We were finally able to determine he was just inside his front door on the floor. The snoring continued to get louder and louder and all we could do was laugh. He apparently made it just inside his front door and then he slid down and passed out. We continued to try and wake him up as the cab driver was pretty irate. The next thing we knew, we heard a really strange noise in addition to the snoring. It sounded like some weird alien creature and at this point we could justify forcing entry into his home to check on his well being because we were concerned for his health. We asked the fire department to be dispatched and cleared the entry with a supervisor who approved it.

Once the fire department and paramedics arrived, they used their halligan to pry the door open. We pushed our way inside, causing the guy to slide across his floor. We couldn't help but laugh. He was out cold. His stomach was extraordinarily extended and he looked as though he was nine months pregnant. The noises we heard were coming from his stomach and they were ugly. I recall wondering if he was about to give birth to the antichrist. It took a while to get the guy to wake up, but once we did, we agreed he needed to get checked at the hospital. We all knew there wasn't much they could do but maybe give him an IV and time to sober up, but if we left him there and he aspirated on his own vomit, we were all liable. No one wanted that.

When I finally got home as the sun was coming up, I told my mom all of the crazy stuff I had seen. I told her I had figured out what I wanted to do with my life. This seemed like the closest I would get to regaining the camaraderie I missed so much from the Navy.

Mom was not happy about my decision. She said I was going to regret it. She never liked my uncle the homicide detective very much so she had a sour attitude about cops too. I promised her I would never become one of "those cops" because of the very reasons I had been treated in the past, but I learned that there were some really cool ones out there doing the job for the right reasons. I told her it was just a matter of getting to know them. I promised myself when I put on the uniform I would never be a hypocrite and would remain as honest and ethical as I had always been. Looking back I realize that's what was missing from the timeshare business. I had been in sales long enough and it was eating at me like a cancer.

Coming in for the five way star for some relative work

October 29, 1998 - The day Senator John Glenn returned to space. What an incredible day!!

Twelve

❧

A New Beginning

Central Florida, 1999

To make sure I really wanted to go into law enforcement, I continued to do several more ride-alongs with Stacy. It never got old, each shift was a new adventure. This was a really nice breath of fresh air from the timeshare industry where I had to give the same sales pitch over and over and over again. It's hard to stay enthusiastic about something once you've begun to feel like a broken record.

Law enforcement was different. No two calls were the same, even if they were at the same house with the same dysfunctional people arguing for the thousandth time, each call was different from the last. I liked the challenge of thinking on my feet and helping people solve their problems. It didn't always work, but the opportunity was there to make the world a better place and help someone even if they didn't know they needed it.

Once I was certain this was the path I was going to take, I had to start the application process. The written application and basic entry test were only the beginning. There was a background check, polygraph test, credit check, physical fitness test, psychological exam and physical exam. Then there was the oral review board. I started the process in the summer of 1999 and it took approximately six months. Two academy classes were set to start in January, one full-time, one part time. Full-time sponsored selectees received a paid a salary while they went through the academy and were guaranteed a job when they graduated. Part-time sponsored employees were guaranteed to be sworn in as reserve officers (volunteers), but the academy was paid for along with uniforms and books. The agency I wanted to work for was on a hiring freeze at the time and only a few were selected to go to the full-time academy. They were all minorities with college degrees and apparently, being a female veteran didn't hold any weight.

When our academy class started, I was one of four females out of a class of 30. I was the only female veteran, there was a former housewife, a waitress and an LPN. We all got along great even though we were quite different from each other. There were several civilian employees from local agencies there who made good money as civilians but wanted to be sworn on the side. There were a few that none of us could quite understand why they were there and none of us saw them ever getting hired, but crazy things happen. Then there were the guys who were there trying to prove something to themselves. No one really got along with them, everything was a competition and it was annoying.

Jessica, the waitress, literally had a photographic memory. She aced every test with a perfect score and she never studied. I studied every night and just tried to stay in the top 5%. I wanted to prove to myself what I could accomplish academically when I really applied myself. Something I had never done before. I managed to maintain an A average and finished in the top 5% which made me really happy.

The academy was fun and we all grew close to each other. I hated PT, just like I did when I was in the Navy, but I wasn't the only one. Not to mention I was the oldest female in the class and definitely couldn't keep up with the young ones fresh out of the military and college. Fortunately, there was no pass or fail physically, you just had to keep up and not fall behind.

Then it came time for the range. As usual, I didn't say much about my background other than I had served. I never talked about my time working with SEALs and I never bragged about my firearm skills because everyone has a bad day now and then and talking crap will catch up with you. Nonetheless, the pressure was on in my own head. This was the area I really wanted to shine. There were several guys from all branches of the military and I was the sleeper. All they knew was that I was in the Navy in an administrative position. They had no idea I had been taught to shoot by the best our United States Military had to offer and I didn't want to let myself or my mentors down.

As the firearms section progressed, it was becoming clear to the instructors that I had skills. They started making comments and putting the pressure on. Two guys were none too happy about that. One was a jar-head and the other a civilian gun nut. He made everyone nervous because he had a hot temper and something just seemed off about him.

By the time the instructors started keeping score towards the end of the firearms block, it was down to myself and Gary, the civilian. I would out shoot him on every course of fire, but only by one or two rounds, so it was close.

Finally the last day came and Gary choked. I remembered everything the Team Guys taught me, "slow is fast." I remembered my breathing, my trigger pull and I remained focused. I aced every course of fire earning the "Top Shooter" Award for graduation. I was so incredibly proud of myself.

The law enforcement driving section was a blast. Everyone loved it. Once we would get done qualifying on the courses, we would play on the skid pad to see who could do the most 360's.

Several of our classmates dropped along the way. Either due to academic issues by failing a test or life would get in the way forcing some to leave. The class got smaller and smaller with each passing month.

Once we graduated after nine months of instruction, those of us sponsored were sworn in at graduation with our perspective agencies. I received several awards including Top Shooter which I was the most proud of. To my surprise, several family and friends attended and none were more proud of me than my mom. Some of my classmates were still trying to get job offers. Gary never did and we all wondered why. Could it have been those weird vibes he put off making everyone uncomfortable? Exactly one year to the day we graduated, Gary called his girlfriend to tell her goodbye. She called 911 and when the police arrived to check on him, they heard a gunshot. Shortly thereafter they heard another one. They set up a perimeter around the house and tried to talk him out, but it was too late, he had blown his head off with a shotgun. He left behind a note addressed to the police department blaming them for not hiring him. As it turned out, Gary had applied with several agencies but rumor had it, he couldn't pass a psych eval. I remember feeling really bad for him because I couldn't understand how someone could take their own life over something like not getting a job. I also felt relief, he hated me in the academy and he really hated me for beating him out of that award. I suppose my guardian angel was working overtime and I didn't know it.

I was told that when I graduated from the academy, I needed to complete the Field Training program where I had to work alongside a training officer to learn the ropes. Most reserve officers held full-time jobs while they went through the process and it took them a very long time to complete the program since they could only get out on weekends. I, on the other hand, had been living off my GI Bill and un-employment while going through the academy. Once I graduated and exhausted my funding, I started living on my savings. I moved in with my mom to save money. Somehow on her disability she was collecting due to her declining health, my savings, and the sale of all my skydiving equipment, we made it through. I completed the training program in 12 weeks, just like a full-time officer would, only I wasn't getting paid.

I made quite an impression on the watch commander and sergeant I had been working for taking on every challenge they threw at me and performed well under stress. I was working every off duty detail I could get my hands on just to make money while working patrol for free. By the time I put in my 40 hour work week and my 30 hours of off duty, the maximum allowed, I was eating, breathing and living the uniform. I had no life, didn't go out and didn't have any time to go out. I was just trying to make ends meet.

Then one night, after working for free for nine months, I was talking to the midnight watch commander. He asked me what my status was on getting hired. I told him I didn't know and that the reserve unit commander just kept telling me that there were several other reserves in line ahead of me. My package appeared to be going nowhere. I expressed my concern regarding money and explained to him that I might have to apply with another agency if ours didn't hire me soon. He said he had never seen anyone work so hard for a job and that the time had come to make things happen.

The lieutenant told me there would be a quarterly awards ceremony that Friday and where it would be held. He told me to show up in my dress uniform, sit in the back and wait for him to call me at the end when everyone had left. He went on to explain that the "Big Boss Man" would always be the last to leave because he liked to hang around and talk to folks after the awards had been given out. I was so nervous about the idea because it was THE MAN, himself, that told me the agency would hire me once I completed the Field Training Officer program. At this point though, I had nothing to lose and everything to gain.

I did just as the lieutenant told me. I showed up in my dress uniform and sat in the back of the church where the ceremony had taken place. I waited until he called me to the front when the "Boss Man" approached him to say hello. As the lieutenant began to introduce me, the boss beat him to the punch and addressed me by my first name. He had an incredible memory. At the time, there were approximately 1500 sworn officers in the agency and he could address most, if not all by name on sight. The lieutenant went on to explain why he needed to hire me as soon as possible before I went to another agency. He told him I'd be a real asset and that he had never seen anyone work so hard for a job.

The boss reminded me of what he told me during the academy, that I needed to complete FTO first. I told him I had, several months ago. He seemed surprised and asked me how long it took me. I told him I completed it in 12 weeks, the same as a full-time officer. I went on to explain that I had been living off my savings and trying to make ends meet with off duty work while continuing to work 40 hours a week for him for free. He was stunned and the LT just looked at him as though to say, "Hire her already." The boss asked me what human resources had been telling me and I told him "nothing, they are still waiting for my file." I explained what the reserve lieutenant had said about others who had been waiting longer than me, but those individuals hadn't completed FTO yet and had been working on it for over a year. Then he looked at the LT and said, "Get her file and have it on my desk Monday." By the following Wednesday, I had a formal job offer and would be starting within two weeks. Finally, after almost two years from the time I applied, I was finally getting hired on as a full time law enforcement officer.

I didn't waste any time finding an apartment of my own as a courtesy officer. It's what most new cops do when they first start out because they typically can't afford to buy a house yet, but they can get either free rent or half rent in an apartment complex in return for security. I did that for a few years until I burned out and started wanting a home of my own. I was tired of moving once a year because apartment managements would change and contracts would end. It was exhausting and I was tired of dealing with the same old noise complaints night after night because someone walked too heavy or talked too loud. Apartment buildings had thin walls and there was nothing I could do about it. When people call the police, they don't want to hear that you can't do anything. Sometimes we can, if there is a law in place, sometimes we can't. If a person is unhappy with their situation, they should do something to change it. Sadly many people don't get that, they want other people to fix their problems instead of doing it for themselves. God forbid you try to suggest that to them and then they get angry.

Not long after getting hired, I responded to my first shooting. I was only a block away when the call went out. I was there in seconds and my backup was at least 10 minutes away. My heart was pounding because the dispatcher was still working to get information to put out over the radio. I didn't know what my victim looked like, where my victim was, what the bad guy looked like or where he went, if anywhere. I was driving into the parking lot with little to no information. Soon I saw several citizens standing around a guy laying on the ground. They were all pointing to the south in an effort to tell me which way the bad guy went. I had no suspect description so I didn't know what I would be looking for. I figured I could gather that information while I stayed with the victim and put it out over the radio for other officers to look. Meanwhile, there lies this guy, in the middle of a grocery store parking lot, with about seven bullet holes in his chest and stomach. Surprisingly he was conscious and breathing. I kept asking him who did this to him and he wouldn't answer me. All he would say was, "That's OK, they gonna get theirs." So basically this guy is laying there, possibly dying and he's not willing to help police get the person who did this to him? Yep! Welcome to the world of drug dealers and gang bangers. No one, including victims, knows anything. It's exhausting, because if he dies, it's a homicide and dead victims can't choose whether or not they want to press charges. My job was to render aid and get info, but this guy was already hardened and clearly had already done prison time judging by the tattoos covering his body. Nothing against tattoos of course, but there is a difference between artistic tattoos and thug life tattoos. They called in a medical helicopter and flew him out. Amazingly he survived and the investigation went nowhere because he refused to cooperate with law enforcement.

About a month later, I responded to my first drowning. It sucked and it would affect me much more than I realized at the time. A family from overseas was visiting on vacation and rented a vacation home with a pool. The family had spent the day barbecuing and swimming. Later that night, the kids were in bed and mom went to take a bath while dad ran to the store. About an hour later, dad returned and did a bed check of the children before he went to bed. Their two year old boy wasn't in his bed and the dad noticed the sliding door open to the pool area. Dad jumped in the pool and pulled his son out while mom called 911. My partner and I arrived at the same time and took over CPR from Dad. Minutes later the paramedics arrived and didn't even attempt CPR at the scene, they scooped the baby up, ran him to the ambulance and took off.

My partner stayed at the house to preserve the scene for investigators and I went to the hospital to interview the family while we all waited to hear from doctors of the baby's outcome. It seemed like forever and I felt so horrible trying to gather information from the family. They sat quietly in shock and answered basic questions from me regarding the timeline of events that night. Eventually, after about an hour of waiting, I noticed a nurse pass by fighting to hold her tears back. Moments later a doctor emerged from the hallway and entered the private waiting room we had been placed in. The doctor sat down calmly, held the mother's hands, and broke the news to her that her beautiful baby boy was gone. I sat in awe as she and her husband quietly held each other and cried ever so softly. I think they were truly in shock and what was happening just wasn't sinking in. It was heartbreaking. I was also surprised to see how all of the medical staff were clearly emotionally affected by the boy's death. I thought based on their line of work, they would be used to this sort of thing by now and numb to it. Then I questioned myself, maybe it was me? Maybe I was a cold hearted person that I wasn't feeling the need to cry over the loss of a baby.

After I finished that report, it was back to work and to all of the other calls holding. I met with my partner later that night and we asked each other our thoughts on the whole matter. We both agreed it was just a tragic accident that maybe could have been prevented had there been child locks on the sliding glass door.

The next morning I got home from my shift exhausted. I went to bed running the case over and over in my head. Hoping I wrote a good report. Angry with the homeowner that rented this home to vacationing families from abroad and not putting something in place to protect small children. I think I was more angry than sad for sure.

When I woke up that afternoon, it all seemed like a blur. I got up, got in the shower as I normally would and for some reason, that was when it hit me. I fell to my knees in that shower and sobbed for that little boy. Why now? Why twelve hours later after I'd slept? I called my mom sobbing and told her what happened. I was confused about why I was just now crying when all the hospital workers cried immediately. She told me that everyone is different when it comes to dealing with death. There is no normal way to respond. Some become highly emotional immediately, others take longer to process. I was in full blown work mode at the time of the incident so I compartmentalized the trauma of the event and ignored it until the job was done. Then for some reason, in that shower, it all hit me and I fell apart.

Not every call was violent or tragic though. There were funny calls too. Once I was sexually molested by an English bulldog who was obsessed with my leg as I tried to comfort his mom because her husband had just died. As my partner and I tried to refrain from laughing at the dog over his stubbornness and obsession, the lady put him behind a barrier designed to keep him away. The next thing we knew, this dog ran head first through that barrier and tackled my leg with everything he had. As I politely pushed him away and reassured the woman it was OK, that I loved dogs, she continued to apologize. I didn't dare look at my partner or we would have both lost our composure and burst into laughter. Finally the woman herself started laughing and we were so grateful because we couldn't hold ours back for another second. I got teased about that one for a long time.

Then there were the perverts. We got a call about a guy masturbating outside of a popular discount dress store. He was watching all the "purdy ladies" as they came and went from the store, according to him. When I ran him through our system to see if he had any warrants, I discovered he had several restraining orders from another county where a major State University was located. It seemed he had run out of places to "watch purdy ladies" so he drove four counties to ours to give it a try. We had enough witnesses, one of which was a minor, to write statements about what they witnessed him doing with himself. So I loaded him up in the car and took him to jail. Unfortunately for me, there was a lot of traffic getting there and I was stuck with this freak in my backseat who clearly was very lonely and desperate for conversation with a woman. I tried everything to shut him up. I tried ignoring him, turning my music up, closing the window between the front seat and the back. All he did was talk louder and refuse to let me ignore him. Just when I thought things couldn't get worse, he said the creepiest thing I'd ever heard an arrestee say, "You sure are purdy... you married?" He barely completed his sentence when I quickly replied, "Yep! Sure am! Got 10 kids! Now shut up!" I felt so gross and I couldn't get this guy out of my car fast enough!

Over the next few years, I worked in the most violent areas in the county. My zone partner and I had just received an award for catching a guy who had beaten a WWII veteran as he was trying to protect his wife from having her purse stolen. I spotted the guy running from the scene and confronted him but he ran. I gave chase on foot through a very busy intersection during rush hour. It was as though we were moving in slow motion and somehow all the traffic magically stopped for us. I wasn't losing him, but not being a very fast runner, I wasn't gaining on him either. Then Kev showed up and cut the guy off with his car. Kev tackled him to the ground and together we tried to put the guy in handcuffs. Unfortunately the crackhead was high as a kite which made him extremely strong. It was like trying to wrangle a bull. Kev applied the taser a few times but wasn't able to make contact as the guy was thrashing about and tossing me like a rag doll. Finally two more officers showed up and between the four of us, we managed to get his hands out from underneath him and behind his back to cuff. We tried every pain compliance pressure point we had been taught, nothing was working on this guy. He was numb to the pain because of the drugs in his system.

Two years later that asshole sued Kev and me. It took five years to go to trial for excessive use of force. It went to federal court because he was alleging it was a civil rights violation because we were white and he was black. After all that worrying, stress and God only knows how many tax dollars to defend us, we went to trial. It took five days to try our case and it took a jury 15 minutes to find Kev and me NOT GUILTY! I'll never forget the jury foreman approaching us when it was all said and done and apologizing to us for what we had gone through because of this guy. Later we found out he had been released from prison only months before that incident happened. He managed to get an early release after being sentenced for second degree murder. Now he was back in prison for robbery and battery on the elderly. A real winner he was.

That's the thing people don't understand about cops. Not only are we forced to make life and death, split second decisions, but we also have to deal with the scrutiny of the media, public and now our leadership second guessing every little thing we do while having the luxury of hindsight which is something we don't have when the shit hits the fan. We do the very best we can with the information we've got in those critical seconds. Do we make mistakes? Sometimes we do. Often it's because we received bad information which led us to make the decision we were forced to make at that time. We don't make decisions out of hate or malicious intent. We are constantly trained to go home, as we should. We hear about cops getting killed nationwide on an almost daily basis and we are keenly aware of how much of the public hates us. The problem is, people don't walk around with a sign indicating to us if they are good or bad. After a while, we become so hypervigilant we just automatically assume everyone is bad until we know otherwise. Trust me, we don't like living this way, but it just happens slowly over time.

Imagine if you could sit at a busy intersection and you had the ability to learn everything you needed to know about every stranger in every car around you by merely running their license plate while waiting for the light to turn green. How many do you think would have criminal histories for attempted homicide, armed robbery, burglary, drug trafficking or labels on them for domestic terrorism beliefs or all kinds of other bad things? I promise you this, it's far more than you can imagine. Nonetheless, unless they are doing something at that moment to give you the right to stop them, you can't. You let them go and wonder, "are they up to no good today?" There is a reason cops are jaded and I promise you, we wish we weren't, but it comes with the job. They say ignorance is bliss, I assure you, that is the truth. Personally I still haven't figured out which I would rather be. Based on my life experiences, I can't imagine what it's like to have the peace of mind some get to enjoy because they don't know what I know. I don't suppose I ever will. I wish I had a dime for every civilian friend who has criticized me for my lack of trust in humanity and my paranoia when it comes to safety.

From the very beginning of my career, I have encountered human beings that have been given every opportunity to make good choices and better their lives, but they choose not to. Some may wonder why cops choose to do this job or why we stay. Aside from the desire to one day say, "I'm retired," there are moments in this career that we live for that make all the crap worthwhile.

While I've had many rewarding moments, the one that stands out most for me, was the first. A woman was walking her newborn baby in a stroller when a drug addicted man attempted to rob her. She had left her home without anything of value, no purse, no wallet, no money, not even her cell phone. So what did he take? He took the baby. He was high on crack cocaine and what he planned to do with that baby we would never know. I suppose it made sense to him at that moment for whatever drug induced insane reason. The baby's mother began to scream hysterically as she tried to fight him and stop him. Unfortunately she was no match for him and couldn't. A passerby called 911 for her and everyone responded. There must have been 30 cops around on that perimeter with K9's and a helicopter on the way. I was still in training at the time, so I had my FTO with me. We took up a perimeter position and listened carefully to the radio to see which way the K9 would track. It didn't take long before we realized they were tracking the bad guy towards us. We stood at our patrol car and kept our eyes focused on any movement in the woods. We listened carefully for any sounds that didn't seem right. It was dark, so it was tough to see and the helicopter was overhead searching with their spotlight and infrared camera. The woods were thick which made it tough, but we weren't giving up til we found them. It was also difficult to hear with the helo overhead, but we hoped to hear something that sounded like a baby crying which should have stood out.

Finally as the K9 German shepherds closed in on them, we heard movement. They were about 30 yards from us in the woods. We moved in to arrest the guy and save the baby. I can't even put into words the feeling that overcame me as I held that baby and confirmed that she was OK. Once the baby was checked by the firefighters and paramedics, we returned the baby to her mother whose tears had gone from that of hysteria to happiness. It's those moments that we live for. They are few and far between, but they make all the bullshit worth while.

Then there are the calls that honestly shock even the saltiest of sailors and the toughest of men. While working a really rough area of the hood where pretty much anyone on the street after dark is dealing drugs or selling themselves for drugs, my partner and I noticed a suspicious truck parked off to the side of the street. The truck was newer, top of the line and whomever owned it clearly made good money. Why was it parked in one of the poorest, most violent areas of our city? My partner assumed it was going to be a "John." "Johns" are what we call men who pick up prostitutes. As we approached the vehicle to look inside, we expected to see just that. Instead, we saw something that shocked us both. My partner at the time, JD, was a good ole country boy with a hilarious sense of humor. I loved working with him because we got into all sorts of stuff and I learned so much from him. As we looked into the cab of the truck, there we saw a well dressed businessman receiving a blowjob from an old toothless homeless man who smelled as though he had been living in a dumpster with raw sewage. It was not a pretty sight. The only thing that made it funny was watching JD hurl the second he realized what he was witnessing. Once we broke them up, we interviewed them. The old homeless guy was just trying to make a few bucks to buy himself another six pack. The businessman was married to a wife and had three kids. I can't even begin to describe our disgust. We tore into him and reminded him of all the things that could go wrong as a result of his actions. We reminded him of how many robberies, shootings and stabbings there were in the area and then we reminded him how easy it would be to catch a disease and take it home to his wife. We asked him exactly how he would explain that to her.

JD and I pondered calling her and telling her what her husband was up to, but we didn't want to be responsible for what happened to him when he got home. So we bluffed him, I made a fake call to another officer and acted as though I was talking to his wife. I told the whole story, in detail, right in front of him. Then we sent him home under the assumption that his wife knew what he had done. We never saw him again in the hood and I suppose we never would. I still wonder what happened to him, but we know he didn't off himself and we never got a call to his residence. With some luck, we hoped he learned his lesson and would never do that again.

Probably the funniest and most embarrassing call we had together during training was the time I patted a drug dealer down. We were in the most dangerous neighborhood in the county where every guy on the street was selling drugs. Finding an arrest was like shooting fish in a barrel.... no pun intended. Anyway, JD pointed out this guy in his early twenties on a bicycle and told me to do a "consensual encounter." Basically that meant to stop and ask the guy if I could talk to him. Most times they would either take off running, occasionally they would agree. In this particular case the young man knew I was a rookie, so he probably figured he could talk his way out of trouble so he rolled the dice. During the course of our conversation, I asked him if I could search him and he agreed. I suppose he hoped I would not reach down far enough in his pockets to feel the small zip lock bag he had hidden under his jeans in his boxer shorts. Too bad for him, JD had trained me well and I found what I could tell was a small bag of crack cocaine. Once I could identify it by feel, I could search him. Upon removing the crack from underneath his jeans, I commenced to do a pat down. As I swiped my hand across his waistband for weapons, I felt a hard object against his abdomen held in place by his pants. As I grabbed for the object which I believed was the grip of a handgun, I pulled, but the object wouldn't come loose. I pulled harder... and when I did, the young man went with it. I informed JD I believed he had a gun as I pulled really hard a third time. JD asked the young man what he had there and he replied, "my junk." I didn't understand what that meant, so I asked him, "what junk?" At this point, my heart was racing as I believed he was armed and I was desperately trying to keep his hands under control as I continued to try and pull what I thought was a gun from his waist. The young man then mumbled a little clearer and said, "My dick! You are pulling my dick." Needless to say I jumped back as though I had just been pulling on a nuclear bomb. JD fell over laughing and apologized to the guy for my cluelessness. Judging by his attitude, he didn't seem to care about what I had done and just laughed along with JD. He went to jail that night with the attitude, "You got me tonight, but you won't next time." To these drug dealers it's all a game. They get arrested, bond out, accept a plea with a public defender and then go back to their trade. It's a lot of paperwork for a cop and after a while, it's pretty discouraging to watch the rotating door at the jail.

Looking back over the last twenty years through all the tragedy and shocking experiences, I will say that I've enjoyed having opportunities to make a difference in people's lives. Some have been homeless and I've helped them get help. Some have been finding missing loved ones struggling from mental illness and out on the street. Making those phone calls to their families, letting them know I've found their wife or child and that they are alive is a good feeling. Arresting abusers and convincing victims that there is help for them if they choose to take it is always a challenge, but one worth taking on. I'm always amazed how an abuser can convince someone that they are as good as it's going to get and that they need someone who is so horrible to them.

Probably the most touching day of my career was a day I spent at a theme park. Every year a flight would travel from Canada to Florida full of terminally ill children and their medical teams. They would leave Canada around 1:00 am and arrive in Florida at about 4:00 am. There they would be greeted by countless cops and firefighters from all over Central Florida.

Our job was to give them a police escort so no time was wasted in traffic. We would escort them through the parks and do all the heavy lifting on and off the rides to include all their transportation. I was assigned to identical twins suffering from the same life threatening disease. They were about 10 years old and weighed only about 50 lbs each. As we spent the day at the theme park, we formed a bond and I was in awe of how intelligent they were. That night when we returned them to the airport and loaded all of their belongings, it was time to say good-bye. I truly choked back tears because I knew I wouldn't see them again. It was such an honor to spend that day with them and I felt like a better person for just getting to know them.

"Anything worth doing,
is worth overdoing.
Moderation is for cowards."
~ Shane Patton

Thirteen

❧

September 11, 2001

Most every adult remembers where they were on the morning of September 11th, 2001. I was asleep at a friend's house who I was house sitting for. I was working midnights then, as most rookies do and I had only been in bed for a couple hours.

The phone rang and it was my mom. "Donna, turn on the television." I was groggy, but I heard the urgency in her voice. I sat up in bed, grabbed the remote and the instant the tv came on, there it was, the World Trade Center tower, engulfed in flames and smoke billowing from it. Mom and I sat there on the phone, watching the tragedy unfold together while trying to figure out how a passenger jet could make such a horrific mistake. As we sat in disbelief, we watched the second jet fly into the other tower. It was at that moment I knew we were under attack and our world had changed forever. It was heart wrenching to watch. People were leaping to their deaths from two of the tallest skyscrapers in the world just to escape the heat of the fire. When it seemed things couldn't get any worse, the first tower collapsed. I couldn't comprehend it. How many people were dying before our eyes? Soon thereafter, the second tower collapsed and as I realized how many of my brothers and sisters in blue and red were inside, perishing trying to save others, I just began to cry. What kind of evil son of a bitch does this?

The day became a blur as bad news continued to roll in. It wasn't long before I heard about the plane crashing into the Pentagon, a place near and dear to my heart. A friend told me that my favorite Skipper from Panama was there that day and I couldn't rest until I knew he was OK.

Then news came in of United Flight 93 crashing into rural Pennsylvania. We would later learn that the plane never made it to its intended target thanks to the heroes that died on board while fighting the terrorists who had hijacked it.

No one slept that day, we were all the way in Florida but it felt as though it happened in our backyard. The attack affected everyone, from North to South and East to West. The country came together and all politics aside, we all wanted those responsible for this attack to pay for what they had done. Patriotism was at an all time high and military recruitment went through the roof. Professional athletes were leaving multi-million dollar contracts to join the military to fight terrorism. Many who had left the military returned, especially in the Special Operations communities.

Finally I got word that my old Skipper had survived the Pentagon attack and had helped save lives that day as his office was a short distance away. I was lucky enough to hear it straight from him as I sat in his office at the Pentagon the following spring when I visited DC during National Law Enforcement Memorial Week. Not only is he a phenomenal leader, he's a true hero and I'd follow him anywhere.

In the days, weeks, months and even years that followed that tragic day, everything changed in law enforcement. We were taking reports on things we would have never given a second thought of about before. People reported every little thing that didn't seem right. It was as if the whole country was on high alert and looking for any clue that might stop any further attacks.

I wrote countless reports on everything from mosquito trucks that sprayed pesticides to tourists taking pictures of unusual things throughout our tourist areas and theme parks. It never ended and we were all working overtime for what felt like an eternity.

If the monsters were evil enough to attack us in NY and DC, we all imagined they were evil enough to attack us in our theme parks. The threat was real, no longer a theory and our defense department had been downsized so much from the Clinton Administration, it now had to catch up to fight this new war on terror.

Most everyone that served in the military during the 1990's when Clinton was elected can tell you the entire military felt the squeeze. Military bases had shut down worldwide and stateside. The Navy no longer had three basic training facilities, they now had one. Cities that hosted the closed bases struggled economically because the military personnel were no longer there to support them. It was heartbreaking.

For years during the 90's Clinton's administration had been warned of the threats coming from the Middle East. They knew of Osama Bin Laden and his training camps. Bin Laden had already attacked several US targets overseas and on four separate occasions the US had the opportunity to take him out, but Clinton didn't authorize the operations. There was more than enough evidence and intelligence, according to the 9/11 Commission Report, that an attack was imminent, yet despite the warnings and evidence, Clinton didn't take them seriously. He was too busy messing around with Monica Lewinsky and dodging his wife's temper tantrums.

President Bush had only been in office a year when 9/11 happened and intelligence determined the attack took years to plan. President Bush inherited a weak country thanks to all the budget cuts that affected our National Security. Nonetheless, the President stepped up and took on the challenge of taking these assholes down. It wasn't long before he gave the orders to send our troops to kick some ass. Americans pulled together and supported the brave men and women deploying unlike any other time in my lifetime. Patriotism was at an all time high and everyone wanted Bin Laden's head on a platter. By this time, it wasn't as easy to find him as it was a few years prior. Bin Laden knew we would come hunting for him and he had prepared for it.

Meanwhile for the next decade, as we tried to build our military and national security back up, our troops were getting deployed for long periods of time with little to no breaks in between. The horrors they were witnessing were horrific and it was taking a toll on them. Military suicides began to soar and a thing called Post Traumatic Stress Disorder or PTSD was to blame. The question was why?

Our greatest generation witnessed all sorts of atrocities by the German Nazis and Japanese during WWII but they didn't come back and kill themselves. As badly as our Vietnam Veterans were treated during the 60's and 70's, you didn't hear about PTSD or epidemic suicide rates. So why now? What was different? These are questions that remain for those far more educated than me, however I have my own theories.

Personally, I believe our WWII vets grew up during a time when life was much harder than it is today. My grandpa would speak of not knowing when the family would eat because of the Great Depression. The men and boys would have to go hunting to feed the family. Times were incredibly tough and as a result, it made them tougher. Once my grandfather amputated his own thumb by accident and instead of calling for help, he picked it up, put it in his pocket, drove home to get his checkbook to pay the doctor to sew it back on. Then he went back to work. That's just the kind of man he was. It's the reason his generation was referred to as "The Greatest Generation."

The Vietnam vets were the sons of these men. No doubt raised the way their mothers and fathers were. Our country failed them when they returned and we are still trying to make it up to them. Despite all that, they bonded together through VFW's and other veteran organizations. They got each other through the dark days while the rest of the country forgot to say Thank You.

Today's generations didn't grow up the same way our previous generations did. We had things pretty easy for the most part. President Reagan had fixed our economy, put fear into the hearts of those that wished to harm us and for the most part, my generation had it pretty good. When most of us joined the military, we really didn't expect to go to war. President George H. Bush had been President Reagan's Vice President, and he himself was a WWII fighter pilot. He was a great Commander in Chief and I felt honored to have served under him. The first Gulf War went so smoothly thanks to great military leaders, it ended almost as fast as it began. Years later, under President George W. Bush, hindsight would become 20/20 when we would realize we should have taken out Iraq's leader, Saddam Hussein when we had the chance, but that's an argument for another time.

Needless to say, my generation and those that followed had life pretty easy compared to our parents and grandparents. In the old days, kids were tougher. We got spankings as kids when we messed up, problems at school were settled on the playground during a simple fist fight and kids didn't carry weapons unless they were going hunting. Blood didn't freak people out because most hunted and cleaned their prey.

Then, along the way, someone decided spanking kids was child abuse, guns were bad, eating meat was bad and technology replaced the great outdoors. Younger folks are starting to lose the art of communication because no one talks anymore. They communicate with their thumbs on electronic devices using abbreviations and emojis.

Then they volunteer to serve their country, and God Bless them for it, but mentally, they aren't prepared. They go to a foreign land where they are hated and witness horrific atrocities, not only toward them, but to women and children who live there. The one thing most veterans have in common is that they joined the military to make the world a safer place to live, not only for the U.S., but for all the innocent lives throughout the globe. Imagine signing a blank check for up to and including your life to protect innocent people, only to be told you can't kill a man you see sodomizing a young boy? Your hands are tied, it's against every ounce of your being, you imagine what you would do if that was your child and you're told you can't do anything unless you are fired at first. This is just one of many ugly examples of things our current military men and women have to deal with. Many of them have held their closest friends in their arms as they have died only to come home and struggle with these emotions and memories. Some come back to find out their wives have left them for someone else and taken their children from them. It's utterly heartbreaking what our veterans endure and our country has got to do a better job of taking care of them upon their return.

September 11th changed everything. Getting on an airplane was no longer an easy task. It became a logistical nightmare. Everyone had to be scanned and searched. You could no longer fly with any type of sharp object, to include a nail file. Passengers were limited to how much fluid they could take on board and everything had to be travel size. No drinks of any kind could be taken through security and lines to get through security checks took forever. Then the "Shoe Bomber" made his move which was thankfully thwarted by alert passengers. From then on, travelers had to remove their shoes while going through security which only caused lines to back up even further.

Meanwhile, law enforcement officers were preparing for every type of threat possible. We were getting issued hazmat suits, gas masks, training for mass casualties, getting terrorist intel bulletin after terrorist intel bulletin. If private citizens had any idea of the immense amount of intelligence and information coming our way on a daily basis, public panic would have most likely taken over. As law enforcement officers, we not only had to stay in the loop and stay vigilant, we had to handle the day to day crimes and calls for service as though it was business as usual.

I often get frustrated when I hear citizens accusing cops of being paranoid and untrusting. People literally have no idea what it's like to know too much. Then there are the Monday night armchair quarterbacks who think they can do our job better. I've never walked into a doctor's office and told a doctor how to do his job. I will never understand why a person who has never been a cop thinks they can do our job better.

As the years would go by, I would reflect on those few years after September 11th when Americans not only appreciated our military for going after the bad guys, they also appreciated first responders, both police and firefighters. Everyone promised they "would never forget," but as President George W. Bush's second term came to an end and the economy had begun to struggle as a result of some lending bills signed by Bill Clinton just before he left the White House, politics got ugly as the Democrats fought to get back in the White House.

During the 2008 Presidential Democratic Campaign, most Americans had their eyes on Hillary Clinton, Bill Clinton's wife. Despite his Impeachment and extramarital affairs, she chose to stay with him. Despite his faults, most democrats still loved him. Many believed Hillary stayed with Bill because she was a stronger candidate with him than without him. After hearing of some of the wicked fights they would have between them in the White House due to his behavior, most of us believed it was a marriage of convenience and for the most part, they lived separate lives. I, for one, would love to see a woman in the White House, but not at the risk of our National Security. I had heard too many stories about how Hillary really felt about those that were sworn to protect her with their lives. She treated her Secret Service agents like shit. A friend who was in the Army assigned to the Secret Service as an EOD specialist during the Clinton Administration told me stories of how she would scream at her security team about the K9 bomb detection dogs. She didn't want to see them or smell them and if she did, then they needed to be "fucking killed." She hated anyone or anything in a uniform and had a nasty temper which was later described in a book written by one of her former agents who described a black eye she gave the President during a heated argument.

While the country was worried about Hillary getting elected, a junior senator from Chicago started to make ground in his campaign. Most Americans had never heard of him since he was only a U.S. Senator for three years, and a state senator prior to that. After law school, he worked as a community organizer and honestly had little to no experience in terms of national security, foreign policy or a clue as to how to be a commander in chief. Barack Obama came out of nowhere and got himself elected President of the United States thanks to celebrities like Oprah Winfrey swooning over him. Many voted for him for the sheer fact that he was black and they felt it was time a black man got into the White House. I have no problem with a black man or woman as President of the United States, so long as they are qualified to do the job and have our country's best interest at heart. I wasn't a huge fan of John McCain, who was running on the Republican ballot, because of things I had heard from older veterans. Many who served in Vietnam with McCain hated him and questioned the medals to which he was awarded because of his father's connections in D.C. Most had nothing nice to say about him. I was, however, a huge fan of his running mate, Sarah Palin. Palin had served as Governor of Alaska and those who lived there, loved her. She was outspoken, direct and a true Patriot. I would have loved to have seen what she could have accomplished had she been elected, but the media put a target on her head and didn't treat her or her family fairly. She was a true conservative and made no qualms about it.

After seven years at war with Afghanistan and Iraq fighting the War on Terror, Barack Hussein Obama was elected to take the helm from President Bush. Those of us who knew how much was at risk were very worried. We prayed we were wrong, we hoped he would do a good job, we continued to do our jobs and time would tell what would be written in the history books.

Fourteen

⸎

God Works in Mysterious Ways

Central Florida 2000 - 2005

  Over the course of the first few years working patrol in the highest crime areas of the county, I experienced every type of adrenaline rush one could imagine. Sometimes they were fun, sometimes they were tragic. I'm sure it's the reason the Heart and Lung Bill was passed for first responders. Cops often suffer from heart conditions after years on the job. The stress of losing friends, hoping they go home at the end of their shift to their families, going from hours of boredom to moments of sheer chaos and then the adrenaline dumps that follow take a toll on the heart. Folks don't realize that unless they find a way to balance their jobs and their off time, they probably aren't going to live long after retirement to truly enjoy it. Somehow though, many never learn to find themselves and create a life of their own outside the uniform. It becomes their identity much like those who serve in the military. It's all they know and they don't know how to live life to the fullest. For those that marry and manage to have a healthy fulfilling relationship, it's easier. Much of that depends on having a supportive spouse who is understanding of what they go through. Sadly, like in the military, that is much easier said than done, but some manage to do it.

As for me, I got married to the job early on. I had given up on dating, decided I never wanted children and while I still talked to Zach on a regular basis and saw him when I could, I pretty much worked six to seven days a week between my normal working hours and off duty. A lot of cops make the mistake of depending on off duty money to live on. Then they buy big houses or expensive cars and commit to payments that are beyond their means if they were to rely on just their paycheck. Where they screw themselves is when they get hurt or something happens which inhibits them from working off duty. It's a huge mistake that fortunately, I never made. I worked off duty for fun money. The one thing I still enjoyed doing for fun was traveling. I was always planning my next adventure and putting money aside for it.

During this time I met someone who would become an angel to me a few years later. I was in an extremely high crime area when a full size van pulled out in front of me. I was forced to slam on my brakes to avoid hitting him. I ran the tag of the van and could see the driver had a long history of traffic violations and arrests. I called out the traffic stop and initiated my emergency lights. The driver wouldn't stop, it was as though he didn't know I was there. It was dark outside and my car had every emergency light on with my bright spot light shining directly into his rear view mirror. There was no way he didn't know I was there, he was just ignoring me. I followed him at a normal speed announcing for him to pull over on my PA system. Finally after several blocks, he pulled into a parking spot of the local drug dealer convenience store. He got out of his vehicle and proceeded to walk toward the doors of the store. By this time I had called for a backup and Chris wasn't far away. I yelled to the driver to come to me and bring me his license. He had a strong accent clearly from the Caribbean Islands, long dreadlocks and was in his fifties. Right off the bat he started yelling at me for harassing him. I explained to him I stopped him for pulling out in front of me, which he denied. Chris pulled up and he had a civilian rider with him that night.

She was a sweet older woman in her sixties, not the usual rider, but I noticed she was wearing dark green cargo pants and some sort of green sweater. I didn't get a good look initially because my attention was focused on the driver who was yelling and gaining the attention of all his friends who were sitting outside the convenience store. Before I knew it, they were all yelling at me for harassing the man because of his race. This was common practice meant to intimidate an officer to leave them alone for fear of being called a racist.

Finally the man gave me his license and I proceeded to write him a ticket. Truth is, if he had just been nice and not given me a hard time, I would have let him go with a warning, but the man wouldn't let me get a word in edgewise so I had no choice but to write the ticket.

Once I was done, I explained to the man his options with the ticket. Judging by his history, I'm sure he knew what to do with it, but I had to explain it anyway. He began to throw an absolute fit as I gave it to him. He started yelling things I couldn't understand and doing some weird dance. Then I heard something I did understand, "VooDoo! VooDoo on you to lose your job!" I was speechless because this was a first for me. Now clearly I didn't believe in VooDoo, but to see someone try to put a hex on me was a bit comical. That's when I saw something out of the corner of my eye. It was the rider standing next to Chris doing the Sign Of The Cross, over and over again and she was praying. I looked at her a little confused when she looked at me and said, "Don't you worry honey, I got you covered!" I burst out laughing and so did Chris. The VooDoo man thought she was putting a hex on him and he started to freak out, backing away wiping his arms as though someone had just thrown flames on him. Then he ran away. After everything calmed down, I met the nice lady. It turned out she was a Catholic Nun who was one of our Chaplains and had several family members who were cops. We became incredibly close over the years and she did loads of ride-alongs with me.

Sister Joan LOVED adventure. Once we had to surround a house with an armed bad guy inside and I gave her clear instructions to stay in the car while I jumped a few fences to get to my position. Several minutes later I heard a sound behind me and Sister Joan couldn't stand sitting in the car, she'd attempted to climb the fences to get to me and fell off of one, ripping the rear end right out of her pants. I yelled at her to get back to the car because it was my responsibility to keep her safe. She had no choice but to listen to me and we laughed until we cried at her predicament. I took her to her car so she could go home once the call ended. I promised to keep her secret so she wouldn't get in trouble as long as she promised to listen to me in the future. I adored her and I loved having her ride with me.

After a few years of working patrol, I developed a desire to become a detective and investigate crimes. I really wanted to work sex crimes and child abuse cases because I thought I would get satisfaction in putting people away who committed those sorts of crimes. Then one night, while on patrol, I responded to a routine missing person call. Didn't seem like anything out of the ordinary when I first arrived. It was a three bedroom house with two men who lived there. They told me they had a third roommate who had moved in about a month prior. One of the men I was speaking with was in a wheelchair and dying of AIDS. The other one appeared to be legally blind as his eyeglasses were extraordinarily thick and in order to see anything, he had to get really close to it. Both men were clearly homosexual and open about it. They didn't offer much information about their missing roommate other than he had gone out the door for a walk on Friday night with nothing but his clothes on and never returned. This was a Sunday afternoon and they were beginning to get worried as the neighborhood they lived in was known for drugs and violent crimes.

I asked the gentlemen if their roommate had taken any personal items with him like clothing indicating maybe he was staying somewhere else and they said, "No, he just said he was going for a walk." I asked if he had left a wallet or something that would have his personal information like his date of birth so I could take a report. The blind dude invited me into a room that was considered a common room shared by them. There was no bed, but dressers and places to keep cloth-

COURAGEOUSLY BROKEN ～ 155

ing, etc. The missing guy and the blind dude shared a room and they each had a twin bed. The terminally ill guy had a room to himself.

As the blind guy escorted me into the common room to look for a wallet or something that would help me identify his missing roommate, I scanned the dressers and as I turned around, my knees nearly buckled. I actually stumbled back from the shock of what I was seeing. There they had created a wall that was covered from top to bottom and side to side of child pornography. The boys in the photos ranged in age from approximately 2 to 10 years old. I instantly felt nauseous and could not believe my eyes. As I continued to scan the room, I noticed what appeared to be little boys worn underwear pinned on the walls as though they were trophies. It was horrific...I couldn't get out of there fast enough. I remember trying to pull myself together before I tipped them off. They literally had no concept that what they had just shown me was illegal... not to mention, DISGUSTING! I remember bullshitting them and telling them I would need to give a detective a call about their missing friend to see if maybe there were any reports over the weekend that would help us indicate where he was. They were so grateful that I was "going the extra mile" for them. Little did they know I was calling for a sex crimes detective to respond and help me with this nightmare I had just discovered.

I called my sergeant and told him what I had. Before calling a detective we had to get permission from a supervisor. My sergeant refused to respond, but agreed to authorize the request for a detective.

When the detective arrived, I explained to him the perverts had no idea why he was there and that they thought he was there to help find their friend. It made for the interview to go more smoothly. The detective asked them if they had any computers and they said they did. He asked them if their missing friend used the computers and they explained they each had their own. Then he asked them about the pictures in the spare room and asked them where they got them. They said they had been downloaded from the internet. At this point he explained to them that these photographs were illegal and that he would need to collect their computers. He explained that he could get a warrant for them or they could sign them over. They agreed to sign them over and they were read their Miranda Rights. The terminally ill guy died before the case went to trial, but the blind guy got over 100 years for possession of child pornography and distribution of it. In the end, the computer forensics unit counted over 1000 images on those computers and it took a very long time to cross reference them all with the National Database of Missing and Exploited Children. The creepiest thing I heard that night was when the terminally ill dude said, "I don't understand the problem. I would have given anything if an older gentleman would have shown me the way when I was young." Those words will haunt me for the rest of my life and they play a big part of why I don't trust anyone around kids. Both of them were so kind and nice. The dying dude had played a character at a local theme park until he got sick. Imagine how many children he checked out in a day?

Needless to say, after that case, I went home and drank myself to sleep. I tried to erase the images from my mind because I never wanted to see anything like that again. I also decided that while I wanted to be a detective, sex crimes probably wasn't a good place for me. The last thing I wanted to do was look at images like that on a daily basis. I have the utmost respect for those that do though, because someone has to put the monsters away. In hindsight I realize that being a sex crimes victim myself, probably made it tougher for me but I had suppressed my experience so deep by this point I wasn't aware of how it affected me. Everything was compartmentalized in my mind and as long as the boxes containing my traumas remained locked and away, I could keep going.

Meanwhile, Zach and I had reconnected and continued our relationship. He would make trips to Florida for various reasons and I would find ways to travel to see him. After the child porn case, I needed to see him. His wife and kids were back in Nebraska where they would spend every summer. I drove nine hours to spend a long weekend with him in Slidell, Louisiana. It was probably one of the most amazing weekends we ever had. He took me to a work party where I met all of his coworkers, their wives and girlfriends. I got to see some familiar faces who knew exactly who I was to Zach and he appeared to have no concerns whatsoever about his wife finding out I was there. It was clear we were more than friends and it was as though he hoped she would find out. Later I was told by a mutual friend Zach's wife was a bitch no one could stand and it was clear he was in a miserable marriage. That just motivated me that much more to hang in there and wait for them to divorce.

Throughout the course of my years working in law enforcement, I maintained my view that while men and women were equal, we were different and that's OK. I strongly believe our genetic design allows for one sex to be better at some things and the other to excel in others. I believe in letting the men have their "boys clubs," the same as women are entitled to their own. With that said, I'll never forget about the time I stopped by the grocery store on my way home from work. While standing in the produce section in my uniform, a strange woman approached me and commenced to praise me for something I knew nothing of. She went on and on about how great it was to see women like me doing what I do. Being the smartass I am and how she caught me off guard, I replied, "You mean picking out produce?" She laughed and explained how wonderful it was to see women showing men that we could do anything they could do. Because of my experiences, that struck a nerve with me. It had been a long day and I didn't appreciate her boldness. I remained passionate that while some women, whom I like to call feminazis, were out there trying to prove something, they were putting lives in danger. So I calmly and respectfully said to her, "With respect, I disagree with you. You see, I am a Navy Veteran and I can tell you from first hand experience, we can NOT do everything men can. The women out there preaching that have never walked in a man's shoes nor in mine. I tried it and I learned, special operations are no place for a woman." I then explained to her in law enforcement, there are certain roles where men are better suited for a job and women for others. I concluded with, "I wouldn't want to trust my son or daughter's life in the hands of a woman if they needed to be carried out of a gun fight with bullets flying." Lastly I concluded with, "and the last time I checked, men can't have babies and women can't pee standing up." At that point, the woman stood there with her mouth hanging open and admitted she had never served in the military or in a job where lives depended on her, so she would have to take my word for it. I shook her hand and thanked her for listening to my opinion and we parted ways. A small victory, but I took it.

It was August, 2004 and Florida got slammed by three hurricanes back to back. I had decided to buy my first home and the timing was crazy. As I was trying to buy the house amidst the hurricanes, I got a call from Zach. He had made Chief, which is a huge deal in the Navy. He was going to be in Mississippi for a couple weeks while he attended the training and initiation. He wanted me to come spend time with him while he was there. I put in for a Wednesday and Thursday off which gave me an entire week off, a perk of working a patrol schedule. I was so excited for him and to see him during such a special time. All the plans were made when we got word the third hurricane was headed our way. My supervisor knew how much I was looking forward to my trip so he gave me a heads up that if I left town early, he couldn't recall me, but if I stayed, he would have to cancel my vacation days which would have ruined everything. I called Zach and told him the deal. He called an old mutual friend of ours who was living in Mississippi and arranged for me to stay at this house until Zach could arrive. What Zach didn't know was that I had gone out with Damon a couple times in Panama after my heart got broken. It never went anywhere, Damon got pissed at me because I wouldn't sleep with him. I felt a bit awkward about it but when I found out Damon had a live-in girlfriend, it made me more comfortable.

I left town amongst all the evacuation traffic. What should have been a nine hour drive took me 13 hours, but I finally made it. I hung out with Damon and his girlfriend a couple days until Zach arrived.

Zach and I had the best time that week. We played a brilliant practical joke on his friend who had no idea Zach and I had known each other for years. We led poor Bobby to believe I was just a chick they met in a bar that shacked up with Zach for the week. Poor Bobby thought

I was crazy, mostly because every time Zach walked away I would do my best psycho bitch impersonation. I would say things like, "I'm going to quit my job and move to Virginia with him." When Zach and I finally told Bobby who I was and that he had been punked all week, we laughed until we cried. The poor guy genuinely thought I was batshit crazy.

Zach introduced me to sushi for the first time and he got a good laugh out of watching me try wasabi. I literally thought flames were shooting from my nose. To this day, I love sushi, but I steer clear of the wasabi and just stick to the soy sauce. He took me to New Orleans for the day even though it was the last place he wanted to go, he went for me. After spending the day with him there seeing the sights, I could see why he wasn't a fan. Nonetheless, I checked it off my bucket list and there was no one on earth I would have rather seen it with.

After I got home, I discovered my new home withstood the hurricane just fine. I was a first time homeowner and it was just me and mom. The house was small, two bedrooms and an unfinished enclosed garage I planned to convert into a guest bedroom. It was a nice quiet neighborhood in a decent part of town. I loved my new neighbors and we became close friends very quickly. It was an exciting time.

About eight months later, a vacancy came up in the Robbery Unit which was considered a major case division. Pretty challenging for a four year officer, but I was up for it. I was working the evening shift by this point from 3pm to 2:30am. After my shift would end, I'd work off duty from 3am to 7am and go home to sleep all day. During that time I was studying for the major case position and the interview was fast approaching.

Out of nowhere I got a phone call from Damon. His unit was sending him to Ft Pierce, Florida to induct one of the boats into the Navy SEAL Museum. Damon would be flying into town late on Thursday night and had to be in Ft Pierce early the next morning. He asked how far from the airport I was and I told him about ten minutes. He asked if he could crash at my place instead of a hotel since it would only be for a few hours. After his hospitality in Mississippi, I of course agreed. I explained to him I would be exhausted however since I was working long nights and I'd be sleep deprived that night because of my interview coming up. He said he understood and he was just looking to crash somewhere.

Thursday rolled around and as I predicted, I was exhausted. I had worked for four nights in a row between off duty and patrol and I had been up all day for my interview. Damon's flight was delayed due to bad weather so he finally got in around 11pm. By the time he got to the house with his rental car, it was after midnight. Despite our fatigue, we sat and talked for a while. He was starving so I fixed him something to eat. We were up for a few hours talking about our love lives. I picked his brain about whether he thought Zach would ever leave his wife and he talked to me about his girlfriend and her two kids that he found to be a challenge. Finally it got to the point where I couldn't keep my eyes open and I could barely form a sentence. I showed him to the spare room where I had pulled the sofa bed out for him to sleep. He asked if he could sit up and watch some TV and I told him I didn't care. I went to my room and crashed.

I suppose it was a result of the exhaustion that I didn't wake up when Damon came into my room. As crazy as it sounds, I think it was all the talk about Zach that made me dream about him when Damon began to make moves on me in the bed. If someone told me this story, I don't know that I would believe them because I recognize how crazy it sounds, but as Damon was making moves on me while I slept and dreamt of Zach, I slept through the whole thing. To this day I have no memory of having sex with him. What I do remember is waking up with him on top of me. I got extremely angry and told him to get his shit and go. Just like that time long long ago, I told myself it was just a dream and it never happened. I was ashamed of myself, how could I let that happen? Zach would be crushed if he knew what I had done. I didn't want to deal with it, so I pretended it didn't happen. So long as I didn't tell anyone, no one would ever know.... Until...

Three weeks later I traveled to Brunswick, Georgia to see Zach while he attended a conference. We had the best time hanging out with

COURAGEOUSLY BROKEN ~ 161

old friends and both of us got so drunk we puked all night. Unfortunately for Zach, he had to sit in meetings all day the next morning. I, on the other hand, got to sleep it off while re-hydrating with Gatorade.

It was an awesome couple days and as usual, saying goodbye never got easier. As I began my drive home from Georgia, I had this horrible gut feeling something was wrong. I kept going over everything that happened that weekend and couldn't come up with anything that may have been wrong. Finally I called Zach as I drove home to ask him if we were OK. He said, "yeah, of course. Why wouldn't we be?" I told him I didn't know, but I just had a weird gut feeling and I wanted to make sure.

For the next couple weeks the gut feeling only got stronger. My best friend, Tina, and I were planning a trip to Vegas for my birthday and I told her about this pending feeling of doom that was eating at me. She asked me when my last period was and I told her I didn't know. Zach wasn't able to have kids by this point so I never worried about getting pregnant. She insisted I take a pregnancy test before we left for our trip and I told her she was crazy. She continued to pester me over the next few days and the only reason I bought the stupid test was because it was a generic brand and only $5.00. I figured it was worth five bucks to shut her up.

I drove home from the drugstore and had to pee really badly. I took the test kit with me and quickly took the test. Before I could put the stick down, I saw a little plus sign in the window. I immediately thought to myself, *well that can't be right*. I sat there for 45 minutes until my legs went numb reading the package over and over again trying to figure out if I was reading it correctly. I called Tina and told her the test was positive. She screamed out, "I knew it!" Then she went on and on about how she had prayed I would have a baby and how happy she was for me. I was less than enthusiastic about it and was not happy that she had been praying for this. I told her I didn't want children. Mom had been pregnant seven times and only I survived. I watched what my cousin went through after losing a baby girl three days after she was born. All I knew was that I NEVER wanted to experience that kind of heartache. I knew nothing about children. I had never changed a diaper, babysat or had a successful interaction with a child. Trying to talk to one was torture to me. I was absolutely clueless.

After hanging up with Tina, I found my way to my mother's room where she sat in her corner watching TV as usual. According to her, I was white as a ghost when I said, "You're never going to believe this." To which she replied, "Well, I'm guessing by that stick in your hands, you are pregnant." I thought, *Oh My God, I'm 35 years old, I've just been promoted to detective, I don't know anything about kids, I'm not married and... and... Oh My God.. Who's the father?* My initial thought went to Zach, but we were certain he was unable to have more children. So how was this possible? Then I remembered those couple hours with Damon... Shit! What was I going to do???

I immediately fell to the floor at my mother's feet sobbing hysterically, apologizing to her for my actions. She laughed and was elated. She said, "Are you kidding me? I've been praying for a while you would get to know the joy of being a mother." To which I replied, "You too!!" What is with you and Tina? Don't you think you guys should have run this by me first?" I cried and cried, I wasn't ready for this. I was terrified. This was not the plan, how in the world was I, of all people, going to handle being a single mom?

Needless to say, I went into denial. I convinced myself there had to be some mistake. After all, I bought the cheapest test there. I immediately went back to the store and bought the most expensive pregnancy test sold. I went home and took that test, I sat for 20 minutes staring at it and in the end, it showed neither positive nor negative. Apparently there is a reason you're supposed to use it in the morning. The hormone found in the urine best shows when the urine is concentrated and a person has been holding it for a while. Now what was I going to do?

Mom suggested I call my OBGYN and have them test me. By this time it was pushing 5pm and they were about to close. The nurse told me to come first thing in the morning and she would test me.

I didn't sleep a wink that night, lying in bed, worrying about how this was going to affect my life. I even called into work sick that night, something I never did. Who the hell could work like this? I was a complete basket case.

The next morning I was at the doctor's before anyone else. The nurse giggled at me and said, "let's get you back and get some answers for you." As I walked out of the bathroom and handed her my urine, I took about three steps toward the room I was going to wait in and I heard her say, "Congratulations Mom! It's positive." She said the second she stuck the lab stick in the urine it immediately turned positive based on the levels of HCG in my system. Later I would learn that those same HCG levels were to blame for morning sickness. As it turned out, that "gut feeling" I'd been complaining about the past couple weeks was "morning sickness" that would last all day. Little did I know they would continue to get worse over the next several weeks making me miserable.

After leaving the doctor's office and calling mom to tell her the results, I had one more person I needed to tell. I was terrified of what he was going to say. He was the love of my life, my best friend and I couldn't imagine my life without him in it. How in the hell was I going to tell him I was pregnant?

I called Zach at work and the second he heard my voice, he knew something was seriously wrong. I told him I needed to talk to him and asked him to get some privacy so he could talk uninterrupted. He did. At this point I was hysterically crying, begging him not to hate me and told him no matter what, to promise me nothing could come between us and that he'd always be my best friend. He replied, "Yes! Of course! Now you're scaring me so please tell me what is wrong!" I just kept begging him not to hate me and he promised there was nothing I could tell him to cause him to hate me. I took a deep breath and I just said it, "Zach, I'm pregnant." There was this long silent pause, I assume he was trying to process what I said. I'm sure he was trying to figure out how he could impregnate me when doctors fixed that a few years prior and as I realized what he was thinking, I said, "Don't worry, it's not yours." That's when I heard a sigh of relief and he immediately congratulated me. He told me he always felt bad that I wouldn't get to experience the joys of being a mom. At this point I was hyperventilating, expressing my fears and doubts. He was so sweet and so calm as he told me what a great mom I was going to be and how he would always be there if I needed anything. Once I calmed down, I asked him if he wanted to know who the father was, to which he replied "Nope!" I sat there awkwardly thinking *he's going to find out eventually and then what am I going to do?* Then he remembered Damon had visited while passing through town and said, "Unless I know the guy." That's when I started crying hysterically again. Repeatedly saying, "I'm sorry. I'm sorry, I'm so so sorry." He said, "It's Damon isn't it?" I said "yes" and continued to bawl. Zach uttered the words, "That son of a bitch! I always knew he was a snake in the grass!" At that point I made an attempt to explain to him how it happened. How exhausted I was when I went to bed and how I was dreaming about him only to wake up and find Damon in my bed. I told him I got angry and kicked him out and hadn't spoken to him since. Zach asked me if I was going to tell him and I told him I hadn't thought that far. I was more afraid of telling him (Zach) because I was scared that I would lose him. He promised me that was never going to happen and he confirmed our plans for July in Virginia Beach were still on.

Once I got that part over with I calmed down. I wasn't sure how I was going to raise a baby on my own, but my mom and Tina had convinced me I wouldn't be alone and I would have plenty of help. I talked to Mom about calling Damon to tell him the news and we both agreed that if he brought up abortion, I was going to shut him down immediately. We agreed for the child's sake he could be as involved as he wanted, so long as he made himself available to answer any questions I might have regarding family, medical history, etc.

Mom and I sat down on the couch and I made the call. She sat silently for moral support. Neither of us had any idea how this call was going to go. I rehearsed this big speech and conversation in my head and no matter how I practiced it, I struggled with getting to the point.

It was early afternoon and I had no idea what Damon was doing other than he was probably at work. To my surprise he answered the phone rather quickly and very cheerfully. As though he was happy to hear from me. My tone was very flat, I asked him if he could talk privately and he said, "Yeah, kinda, I'm in the back of a Humvee driving across Fort Hood with some of my guys." I told him to give me a call back later when he had some privacy and he insisted he could talk then. I knew it was going to be awkward, but he really didn't give me much choice. So I started out with, "Ok, so you remember when you were here last month?" To which he replied, "Yeah, why? What's up?" I sat there for a moment trying to figure out a gentle way to say what I needed to say, but I realized there really wasn't a way to say it other than just come out with it. After a long pause, I said, "Damon, I'm pregnant" to which there was silence. I waited another moment and followed up with "and it's yours." Still silence. After what seemed like forever of total silence, I said, "OK then, like I said, you probably need some privacy to have this conversation so why don't you call me back later." After a short period of continued silence, I hung up.

I looked at mom and we kinda laughed. I could only imagine the look on his face in the back of that Humvee. I wondered if I had been on speaker phone and all the guys heard it. Mom and I figured we wouldn't hear from him for a while so we just started talking.

To our surprise, the phone rang a few minutes later and it was Damon. Apparently he wasn't feeling well and the guys had to pull over so he could get some air. Later I would find out the real reason was that he needed to puke. As stressful as that situation was, I still laugh about it to this day. There he was, the big tough Senior Chief, puking his guts out along a roadside in the middle of the desert with all his guys watching.

When Damon called back he took ownership of the pregnancy. He said he knew he had messed up and that what he had done was wrong. I told him he could be as involved as he wanted to be for the sake of the child and that I didn't want to come between him and his girlfriend. That's when he told me they had just gotten married. Needless to say he was scared to death of what she was going to do. He imagined going home to find an empty house because he thought she was going to wipe him out. I told him not to say a word for a while until we knew that the pregnancy was going to stick. I told him about the high rate of infant mortality in my family and that I wasn't telling anyone outside my closest family and friends until the second trimester. He said he couldn't keep a secret that long and that it didn't feel right. Against my advice, he called her that night and told her. To everyone's surprise, she didn't leave him. I figured that she stayed for the money and military benefits. She was a single mom with two boys and she didn't work. She was a bit of a wild child and much younger than Damon. Her lifestyle eventually caught up to Damon by costing him his career. Allegedly she was having an inappropriate relationship with the wife of one of Damon's men and he apparently was charged with fraternization, a classic "guilty by association" example. Fraternization has never been tolerated in the Navy, but since he had served honorably for 20 years, the Navy allowed him to retire honorably so he could collect his benefits. Many believed he was a top pick for Master Chief on the next promotion cycle.

Later that night Damon called me and told me his wife forbade him to ever speak to me again. He told me to ask him whatever I needed to know at that moment because if he wanted his marriage to work, he couldn't have any communication with me. He instructed me to direct any questions or concerns to his wife moving forward, something I refused to do.

My mom was writing down questions as fast as she could and handing them to me, his social security number, place of birth, blood type, family medical history, physical address and everything else we thought we might possibly need in the future. It was a strained and awkward conversation and his tone of voice had completely changed from just a few hours prior. It was clear she had put doubt in his head that he was the father and it appeared as though he suspected I was looking to take advantage of him because of his military status. Clearly she was trying to convince him I was capable of something she, herself, had already done. I was offended since I was doing fine on my own and would never do such a thing to someone, but since I had no emotional ties to him, I let it go.

Mom, Zach and Tina assured me everything would be fine and I wouldn't be alone. I had a good job and financially, I could handle it. The thought of abortion never even crossed my mind. I have always been adamantly against abortion. I believe that every life is created for a reason, even though we may not know it. I suppose if I was young and didn't have a good career I would have turned toward adoption, but I knew good and well, at the age of 35, this baby was conceived for a reason and honestly, when I looked at the odds of me actually getting pregnant under these circumstances, I was sure of one thing...God had a plan for me, it wasn't my plan, but it was His.

Fifteen

Not Exactly How I Planned This

I was 10 weeks pregnant and still working in patrol waiting for my transfer date to the criminal investigations division when I noticed a disabled car blocking an intersection. I hadn't told anyone at work I was pregnant and was trying my best to keep it a secret. The car was blocking traffic and it needed to be moved. I took it upon myself to physically push it out of the intersection and call help for the motorist. Shortly after leaving, I felt something was off. I pulled into a fast food restaurant to use the bathroom and discovered I was bleeding heavily. I began to panic and called my mom. She told me I would have to go to the hospital and I could either call someone to come get me or call for an ambulance. Either way I had to tell my sergeant what was going on so he would know I had to leave my patrol car and needed someone to pick up my weapons.

I sat on the toilet freaking out and called him. I was crying as I tried to explain to him I was sorry, but I needed to go to the hospital because I thought I was having a miscarriage. He kinda freaked out considering he didn't even know I was pregnant. He responded immediately and offered to take me to the hospital himself. He arranged for someone to get my car and weapons and I begged him to keep it a secret. He informed me that he had to tell the watch commander so he would know where we were. As luck would have it, the watch commander that night was the same lieutenant responsible for getting me hired.

I sat in the backseat of his car talking to my mom in detail regarding my situation. I could tell the sergeant was extremely uncomfortable as he drove as fast as he could to the hospital trying his best to tune me out. After we arrived at the hospital, they took me straight back since I was in my uniform. The lieutenant showed up for moral support and was incredibly sweet as he lectured me for pushing a car while pregnant. The ultrasound tech showed up and checked my belly, there she pointed out a tiny little peanut with a heart beat. That was the moment I realized I really did have a baby growing inside me. Seeing that little heartbeat captured me and I instantly fell in love. I started crying and was quick to show the picture they had given me to my lieutenant when he came back into the room. He promised not to tell anyone and assured me I could tell the agency when I was ready. Soon thereafter, the doctor showed up to examine me and read the ultrasound. She determined I had strained myself and had possibly torn the placenta causing me to bleed. She ordered me to bed rest for a few days until I could follow up with my OBGYN. I took it easy all weekend and returned to work the following week.

Soon thereafter I finally got transferred to the criminal investigations division. I was assigned to an officer who clearly was not happy about training me. He gave me a hard time from the moment I showed up. Later I found out he had a friend that interviewed for the same position but I was selected instead. He clearly was holding that against me and he didn't even know me. He treated me as though I was an idiot and spoke down to me. If there were ever a perfect example of a chauvinistic pig, it was him. No matter how hard I tried and did as he asked, he found something wrong and criticized me every step of the way. I was really struggling with morning sickness by this point and doing my best to hide it. I would bring light snacks to work because I couldn't eat much due to the nausea. He would make comments like "Every time I turn around you're eating, this is a unit where you will go days without sleep and little to eat." He tried his best to discourage me from doing the job.

Then one day he called me and told me to report to the mall immediately due to an attempted carjacking. The scene was contained and he was already there with several other officers. I knew I needed to eat, so I drove through a fast food drive thru and grabbed a cheeseburger. When I arrived at the scene and he saw that I got something to eat, he flipped out yelling at me for getting food. Then he demanded I get out of the car in the pouring rain just to confirm a VIN number that I already could read off of my computer.

By this time I was furious with how he was treating me, so I said, "Fine! I'll get out of this car, stand in the pouring rain in a white shirt! I'm sure that will look professional." The jerk stood there with his giant umbrella and wouldn't offer to let me use it. He was intentionally trying to make me miserable and he was doing a good job.

After we were done with that call and I was driving home exhausted and frustrated, I called him. I told him I was only telling him a secret because I needed him to understand my situation and where I was coming from. He was married, had three kids, I thought he would understand. When I told him I was 11 weeks pregnant, he freaked out and said I shouldn't be doing this job because of the work they do with stakeouts and take-downs. He insisted I report the matter to the agency and I told him that I wasn't ready yet and he was to keep this information to himself.

The next morning when I walked into the office, there he sat in the acting sergeants office talking to him. I didn't even have a chance to put my stuff down when I got called into the office. There I was lectured about how challenging the position was and how a woman in my condition shouldn't take on that kind of responsibility. They then went on to tell me that if they wanted to get rid of someone, they could. I was furious. I had worked so hard for this position and I hadn't even been given a chance.

The asshole told the Corporal that he refused to train a pregnant woman when he didn't even want to train a female in the first place. I just sat there in awe that he had the audacity to say that in front of me. Later I would learn he had already gotten in trouble once for harassing a female elsewhere in the agency. Had I known that at the time, I would have filed a formal complaint, but I already had enough drama on my plate as it was.

Soon thereafter, the Captain called me into his office to talk. He was kind and understanding. He convinced me this new position was extremely demanding and stressful. He told me he had children himself and he promised me that while I may not know it then, putting my health and the baby first would be the best decision I could ever make. He promised me the job would still be there waiting for me when I was ready. Part of me knew he was trying to handle the matter diplomatically before I filed a complaint against the two chauvinistic assholes, the other part of me knew he was right.

I went on "light duty" which basically meant I was working a desk job and answering phone calls from pissed off citizens. I hated it, but I had bigger things on my mind so I just went through the motions.

Then July finally arrived and it was time to head for Virginia Beach. I was just entering my second trimester and barely starting to show a baby bump. Zach and I had made plans to go to the big annual SEAL reunion. I was so excited to see old friends and spend time with Zach. It was a much needed break and getaway from all the stress I had been dealing with.

When Zach and I were at the reunion, several friends kept trying to buy me a beer, but I continued to refuse. I was trying to wear clothing that would hide my baby bump and for the better part of the day, I was doing a good job. Then a good friend of ours who had way too much to drink that day, started showing his ass giving me a hard time about not drinking. One of the women in the group figured out why I wasn't drinking and she leaned in to ask me quietly. I confirmed her suspicions and she giggled and told me not to worry, she would have my back. The next time Stevie started running his mouth about the fact I wasn't drinking, she stepped in and told him to shut up, not everyone was a lush like he was. But that didn't do the trick, he kept on until he got a good reason why I wasn't drinking, I figured the secret was going to come out eventually so I blurted it out, I said, "I'm pregnant you dumbass!" The look on his face was priceless. He had been like a brother to me for ages. His entire drunken demeanor changed as he suddenly decided I needed to be treated with kid gloves. He brought me a chair, insisted I sit down, kept checking on me. All the while he assumed Zach was the father and continued to congratulate us. It was awkward, but I couldn't have asked Zach to handle it any better than he did. He let it go and allowed everyone to believe the baby was his. He treated me like a princess that entire weekend and made sure I was comfortable no matter where or what we were doing.

When it came time to say goodbye, it was unlike any other goodbye we ever had. We had plans to see each other in October, but I knew something was different. THIS goodbye was going to be the last goodbye for a very long time, despite what Zach told me, I just knew. Maybe it was the pregnancy hormones, but regardless, I couldn't turn off the waterworks and I couldn't let go of him as we said goodbye in that airport. I told him, I feel like I'm never going to see you again and it's killing me. We both choked back tears and he promised me I was wrong and we would see each other again.

I made it back to Florida and Zach and I stayed in constant communication. Damon's wife had started to harass me and send me nasty messages which only added more stress. Finally I had enough and our friends encouraged me to hire an attorney to make the harassment stop and represent me in fighting for child support when the baby was born. Damon and his wife had done it to themselves, they forced me to play hardball, which was the last thing I wanted to do, but they had asked for it at this point. All of our friends abandoned him and sided with me. I would get yelled at for being too soft on him, but I suppose they could see the future better than I could when I was just trying to get through each day.

Soon thereafter it came time to undergo some testing for women over the age of 35. By getting an amniocentesis done, doctors could determine if the baby was healthy. As an added benefit I could find out the sex of the baby with 100% certainty. When the nurse prepared me for the test, she used an ultrasound monitor. By this point I was convinced I was having a boy. I had picked a name, nursery theme and colors. When the nurse put the ultrasound wand to my belly, the baby literally put her buttocks right up to the wand and spread her legs. She wanted to make it very clear to me she was a girl and not a boy.

The nurse laughed and said, I never say this, but I'm willing to bet my check that your baby is a girl. The amniocentesis test confirmed it and I cried and cried. I immediately began to worry about how difficult raising a daughter would be and dreaded the day she would be a teenager. My mom got the biggest laugh and promised me I would be glad I was having a girl once I let it sink in. She was right.

When I told Zach I was having a girl, he was thrilled. He had two girls of his own and he told me all the reasons girls were so much fun. I told him I had decided to name the baby after him if it was a boy, but now, since it was a girl, I wasn't sure how to name her. After giving it some thought, I suggested the name Sophia and both Mom and Zach agreed they loved it.

In August, a Category 5 hurricane struck New Orleans, Louisiana. Her name was Katrina and she devastated the Gulf Coast. I gathered as much clothing and donations as I could from friends. I drove them to Tampa to a retired Master Chief who would be making trips to deliver donations to all the Navy families affected in Gulfport, Mississippi. Despite everything that had happened, I wouldn't have wished that kind of nightmare on anyone and I genuinely worried about my brothers who were there.

A few weeks later, in my 24th week of pregnancy, I started having contractions. I was given medicine and sent home to take it easy. Two weeks later, the same thing happened. Eventually I just got used to feeling contractions and unless they were really painful or became regular, I tried not to worry about it.

Then, on October 6th, I was talking to Mom at the house when a tiny bit of clear liquid began to run down my leg. I was 28 weeks pregnant and not quite out of my second trimester. I thought the baby was just stepping on my bladder but mom convinced me to get checked.

Off to the hospital we went again, but this time was different. The nurses did a test which could predict the likelihood of labor starting in the next 24 hours and I tested positive. The next thing I knew, I was being prepped to be moved to a major hospital downtown because if the baby came, this hospital wasn't prepared to care for a baby that premature.

They put me on an IV of magnesium sulfate and I hated it. It made the room feel as though it was 120 degrees and I saw five of everything as though I was looking through a kaleidoscope. As all of this was going on and I was being prepped for the ambulance coming to transport me, my phone rang and I answered it.

It was my uncle, my father's brother. He was calling to lecture me about my father getting old and struggling. I had made it clear to my uncle a few times over the years that I wanted nothing to do with my father, but every time he got tired of being his caregiver, he would call me in an attempt to pass the responsibility.

Needless to say, when I told him I couldn't talk, he got angry demanding to know why. When I told him I was in labor, it was as though he needed a translator. I got agitated with him and handed the phone to my mother. She hadn't spoken to him in years, but she made an exception that day to take the pressure off of me.

When I arrived at the big hospital downtown, they held me in the delivery area and ignored me because my labor wasn't progressing. I was having so much trouble communicating because of the drugs they had me on and I felt as though I was paralyzed. I couldn't move my arms or legs. I just kept asking for water and was told to get it myself.

By this point my mom had gone home and I was in tears. I called her and explained what was happening and she called the hospital and raised all kinds of hell. Soon thereafter I had more nurses than I knew what to do with. They apologized for not understanding my circumstances and they moved me to another ward for mothers who were restricted to bed rest.

The next day a neonatologist came in to visit me and offered to take me on a tour of the Neonatal Intensive Care Unit (NICU). When I saw the tiny babies clinging to life with tubes and wires all over them, I was overwhelmed. The doctor told me that every day I could stay pregnant, I would spare the baby 3-5 days in the NICU. It was then and there I decided I would do whatever it took to stay pregnant so that Sophia wouldn't have to endure what those babies were.

My room was kept extremely cold but I never noticed. I stayed in bed and watched the Health Channel and drove my doctors nuts with medical questions and "what if's" every day. One doctor got so frustrated with me he said he was going to see if they could block that channel from all the pregnant women who do nothing but lay in bed and learn all the things that can possibly go wrong. I loved my doctors, they were confident and kind.

Visitors would come and pamper me. Mom brought me Kool-Aid every day as it was the only liquid I liked during my pregnancy. For some strange reason, I loved things I normally hated and hated things like pizza that I normally loved.

I was worried about running out of vacation and sick time before the baby came and my employer just kept telling me not to worry, they had me covered. I still worried nonetheless, it's what I did best.

On October 19th, I wasn't feeling well. Mom had been there as well as Sister Joan, who had continued to visit me regularly. Having her visit meant the world to me, I just adored her.

The nurses continued to have trouble getting vitals from Sophia since she kept hiding in my belly. They continued to adjust the belly monitors and by the time I was going to bed, I was on an oxygen mask. I felt tired and I didn't know why, I just wanted to sleep.

On October 20th, at 7:00 am, my room lights went on and I opened my eyes to a room full of doctors and nurses. I was hearing things like "prep the O.R. STAT" and I knew something was seriously wrong. I asked what was going on and my doctor told me it was time to have a baby. I argued and told him I still had 10 weeks to go. I was only 29 weeks along. He explained that something wasn't right with Sophia and it was time to take her out because they could take better care of her on the outside than inside when they didn't know what was wrong.

Sixteen

❦

Divine Intervention

I will be the first to admit, while I have never stopped believing that God exists, I will say that over many years, I lost my Faith in Him. After all, I had been through hell and no matter how much I prayed, it seemed like God never answered my prayers. I felt abandoned and unimportant to Him.

Then, when it seemed like my worst fears were coming to fruition as the doctors and nurses scrambled to get me to an operating room, I noticed that I could no longer hear Sophia's heartbeat on the fetal monitor. I began to completely freak out, asking over and over again "Is she going to be OK?" It was as if my words were falling on deaf ears. No one would answer me, most likely because they honestly didn't know. They wheeled me into the prep room to prepare me for surgery and I noticed one nurse look at another nurse with a sorrowful expression and a subtle shake of her head as though to say, *it's too late.*

My body was violently shaking with fear, I was crying, I was calling my mom and everyone I could begging them to come to the hospital. I had truly never experienced fear like I did in those moments. I was shaking so hard, the metal on the hospital bed could be heard rattling.

I couldn't take it anymore, so I grabbed my nurse by the collar, pulled her to me, looked her directly in the eyes and said, "I have delivered enough bad news to people in my career to know how to read your body language and understand your silence while I beg you for answers! Is my baby going to make it or not?!"

She looked at me and said, "We don't know Sweetie, that is why we are going as fast as we can to get her out." It wasn't exactly the answer I was hoping for, but it was an answer. She tried to get me to calm down, but I couldn't, I was terrified.

I knew there was no way my mother was going to make it to the hospital on time due to rush hour traffic and the distance. My neighbor, Carly, offered to drive her but she couldn't get ready fast enough. I called Sister Joan. As it just so happened, she was only a block away and could be there quickly.

The anesthesiologist came to me and introduced herself. She explained to me that they were going to have to do a spinal block on me for surgery since they couldn't give me drugs and needed to block the pain of the Cesarean Section. I said I understood, but I was just too scared to stop the adrenaline flowing through my veins causing me to tremble so violently.

Finally my surgeon showed up and tried to calm my fears. Nothing anyone said to me helped. I was too keenly aware of the fact that I couldn't hear Sophia's heartbeat and they couldn't find it either.

Soon they began to wheel me to the operating room. It was just my nurse, the nurse anesthetist and me, all females as we entered the sterile operating room. That's when a feeling came over me I had never experience before or since, it felt as though someone wrapped me in a warm blanket, hugged me tight and I heard a man's voice say to me, "Everything is going to be alright." I immediately stopped shaking, stopped crying and I felt a Peace overcome me that I never knew was possible. I looked around me to see where the voice came from and there was no one there except us three women. No one had placed a warm blanket on me, we were simply entering the operating room as it happened. It was as though someone literally took all my fears away in an instant. No one had administered any type of drug to me to cause me to hallucinate or calm me down. There truly was no logical explanation for what I heard and felt. It was in that instant I realized God himself had spoken to me when I needed him most. I no longer felt alone and I genuinely believed that everything WAS going to be OK.

I remember the nurse anesthetist even commenting to me what a great job I had done calming down. Shortly thereafter, Sister Joan came walking through the door all dressed up in scrubs. She would be there with me through the surgery and be there to hold my hand as my surgeon would deliver Sophia.

Sister and I said one Hail Mary after another as the surgeon got to work. It was as though one prayer stopped and the next immediately began. I could feel the intensity of the prayers and God's presence in that room and while still scary, I felt my Faith for the first time in many, many years. I remember telling God, "Lord, if you please save her, I will change my ways and I will live my life setting a good example for her." I thought about my relationship with Zach and pondered the sins I had been committing for so long. I knew I needed his friendship and I felt as long as our relationship stayed on a friendship level, God would understand, so I promised him I would stop seeing Zach in the physical sense and maintain a platonic friendship.

I remember Sister Joan telling me step by step what she could see over the giant blue sheet that blocked my vision of what was happening. I could feel pressure as the surgeon reached deep into what felt like my rib cage to grab Sophia and deliver her. Sister Joan exclaimed how beautiful she was as they pulled her out, but I wouldn't be able to see her yet until they revived her. They took her away quickly and began to work on her to get her breathing. Finally we heard a very faint cry and it was the most beautiful sound I'd ever heard. They gave me a quick peek so I could see her before they passed her through the window to the NICU.

Finally my mom arrived and switched places with Sister Joan. Mom was there with me as they sewed me up and we were both so happy I hadn't been alone through the delivery.

All I could think about was how Sophia was doing and wanting to see her. Once I left the operating room I was wheeled into the recovery room where everyone was waiting for me. Tina had driven from the coast to be there, my neighbors, my mom and Sister Joan. Everyone assured me I would be able to see Sophia as soon as possible, but the neonatologists had work to do on her and I had to get stable enough to go see her.

Over the years, one thing I learned about myself was that I was a lightweight when it came to meds and alcohol. Consequently, when the nurses would come to me with pain meds, I refused them. I was adamant that I wanted to be of clear mind when I got to see Sophia and more importantly, if I had to make any major decisions. I wasn't feeling any pain, so as far as I was concerned, I didn't need any pain killers.

About four hours after she was born, I FINALLY got to go see Sophia. Carly offered to wheel me to see her in a wheelchair because I was still numb from the spinal and couldn't walk. When we arrived at the entry door to the NICU, I was taught how to scrub in and was told I had to do that every single time I was going to go through the doors. Even if I left for a minute, I had to scrub in again to return. The purpose of this rule was so no germs of any kind could enter the area because the infants inside were all considered critical and had little to no immune systems. Carly went with me and had the forethought to bring a camera. I was so glad she did as she captured my face the first time I truly got to lay eyes on Sophia. I couldn't hold back my emotions, I just cried as I sat there and stared at her with her tiny little hand that couldn't even reach around my finger. She was so fragile and hooked up to so many wires and tubes.

She wasn't breathing on her own, so they had her on a ventilator which breathed for her. There were tubes coming from her belly button that allowed doctors to check her lab levels every few hours. There were heart monitors connected to her which constantly showed her heart rate and oxygen levels and if her heart stopped, alarms would go off.

I felt so guilty, as though I had done something wrong to cause this. I wanted so badly to fix her, but I couldn't. I went back over my pregnancy and I couldn't figure out what I could have done to prevent this from happening. Doctors assured me I had done everything right and that sometimes these things just happen. Sometimes babies are just in too big of a hurry to come into the world and they don't know what's best for them. They also assured me that preemie babies grow up to be very strong willed and successful because they have that natural ability to fight for themselves. Considering Sophia was the daughter of two veterans, that didn't surprise me.

I spent as much time as I could with her until they made me return to my room. They insisted I take some pain meds before going to bed and I continued to refuse. Mom had called Zach and kept him updated on everything. He was really concerned about me and Sophia and when I was finally able to talk to him, I broke down and cried. I just vented all my fears and concerns to him and in hindsight, I feel so horrible for dumping all that on him because he was 10 hours away and couldn't do a thing to help me. I could hear his concern and regret in his voice. He literally didn't know what to say. He always got quiet when he was deeply concerned about something and this was no different. He did his best trying to convince me everything would be OK, but he knew that was a statement he couldn't back up. I was on an emotional roller coaster at this point seeing her and worrying about her.

The next morning I woke up and when I went to move, I screamed out in pain. I must have hit that nurses call button countless times until someone finally came to my room. I was in agony and it literally hurt to breathe. That's when the nurse got to say "we told you so." My body had literally been cut from side to side, organs moved around, muscles and nerves cut and now that the spinal had worn off and I was no longer numb, the pain had set in. I couldn't stop crying from the pain and all that did was make it worse. I finally agreed to take the pain meds and was told it would take a little while to "get on top of the pain" which is why they suggested taking them ahead of time to prevent it from getting that bad. Lesson learned for sure.

A dear friend from work who was also pregnant and due around the same time I was, called me to check on me. All I did was cry and tell her how much pain I was in. Probably not the best thing to tell someone in the seventh month of pregnancy. Later I would apologize and explain I was out of my head. I also passed on the wisdom of taking the pain meds if and when they are offered. Don't be a stubborn ass like me.

Once the pain meds finally kicked in, I was able to go see Sophia again. While in the NICU, I noticed a baby immediately next to her in another incubator. The father looked very familiar to me and I kept catching him staring at me as though he was trying to figure out where he knew me from. The next day it came to me where I'd seen him before. I had arrested him in the past and I was completely freaked out he would figure out where he knew me from, especially with Sophia's name right there on her medical chart so easily accessible to anyone in the room. I learned the baby was born in critical condition due to the mother being on drugs during her pregnancy, that's when I was able to figure out why I thought I knew him.

I reported my concerns to the nurses and within hours, Sophia was moved to another NICU Unit on another floor. There she had a private room and a window. It made visiting her more private and peaceful, not to mention a lot less stressful.

Later that day I got to meet the "Nipple Nazis." That is what the doctors and some nurses called the lactation nurses. They do a great job convincing mothers that if they don't breastfeed, they're a terrible mother. I learned all about the "liquid gold" medically referred to as colostrum. It's the milk produced first before regular breast milk comes in. Sophia was too premature to know how to eat on her own yet. A baby born at 29 weeks hasn't developed the ability to suck, swallow and breathe, so they have to be fed with a feeding tube. I was given a breast pump and taught to use it. It wasn't easy and it was really painful, but I did my very best to pump every few hours if that's what she needed.

During my pumping lesson I got an unexpected visitor. A former lieutenant of mine walked in to visit me after he heard I had the baby from a good friend of ours. Needless to say, he got an eyeful as he walked in. With everything going on, it was a laugh I seriously needed. He turned 100 shades of red in embarrassment and all I could do was laugh at his expense. What they say is true, after you've had a baby, you have very little modesty. I honestly just didn't care.

He managed to go see Sophia thanks to a nurse who took him with my permission. Later he returned to my room and awkwardly apologized about walking in earlier. I assured him I just didn't care and I was just happy to see him. During my stay in the hospital I had come to feel abandoned by my agency. No one called or checked on me from work. My best friend from work was still in Mississippi helping with the recovery efforts after Hurricane Katrina, so seeing the lieutenant whom I really liked, made me happy.

The next day a few other former coworkers came to visit me and see the baby. None of them knew I had been in the hospital for weeks until word got out I had the baby and she was in critical condition.

On the third day, I finally got to hold Sophia for the very first time, it was only for a matter of seconds, but I cherished them and cried tears of love and joy to finally get to hold her. The nurse was changing out the sheets on her bed and it had to be done quickly so she could be placed right back into the incubator. I was so grateful she let me hold her, even if it was just for a few seconds.

On the fourth day I was discharged. I had been trying to prepare myself for leaving her behind, but nothing could have prepared me for how I felt when I got home and saw the empty crib in my room. I immediately fell to the floor sobbing. My mom was there to hold me and tell me everything would be OK. She told me there was a name for what I was experiencing. It was called "empty arms syndrome," the grief that hits a mother when she returns home after having a baby and that baby isn't there to be held. It's a horrific feeling and I wouldn't wish it on anyone. Even though I knew Sophia was alive, the fear of not knowing when or if she would ever come home was more than I could handle. I was a mess.

To add even more concern, Florida was being hit by a hurricane that night and I worried about Sophia's window and the fact that she was on the top floor of the hospital. I was assured the hospital windows could withhold much higher winds that this hurricane was bringing, nonetheless, it was one more thing to worry about.

That night I must have called the NICU every hour checking on her through the night as I couldn't sleep. They were so patient with me and I suppose I wasn't the first or only mother to do that.

Sophia's neonatologists were amazing. They took their time explaining everything to me in terms I could understand. One of them had been featured on television as one of the best in the country. I loved him.

All of the nurses and respiratory therapists were awesome, except one. On the tenth day in the NICU, I noticed her respiratory therapist (RT) seemed to care more about what celebrities were doing according to a magazine than keeping an eye on Sophia. I remember leaving that night with a bad gut feeling. I made sure the RT acknowledged me as I left so she would pay more attention to Sophia. All of the other nurses and RT's had interacted with me during my visits but not this one.

It was only a 30 minute drive home and I had just walked in the door when my phone rang. It was Sophia's nurse telling me there had been an accident. I immediately drove back to the hospital and the RT was not there. It seems while in the course of doing her job of changing the tape on Sophia's breathing tube, she broke not one, but THREE safety protocols. How do I know this? Sophia's doctor was furious and he told me everything. It was sheer laziness and carelessness. Apparently she had changed the tape by herself when it was supposed to be a two person job. One to hold the infant's hands out of the way. Two she had failed to pre-cut the tape outside the incubator thereby breaking the third protocol which was to NEVER put anything sharp inside near the baby. As a result, the RT had literally severed Sophia's right index finger so badly, a plastic surgeon had to be called in to rebuild her finger. As a result, she had also introduced germs to Sophia by lacerating her fingers with a non sterile metal object. This caused her to be put on antibiotics to fight whatever infection may come, which also caused her to get diarrhea which made her lose weight. While 3 pounds was considered big for a 29 week preemie, 3 pounds is also dangerously small. Sophia couldn't afford to lose any weight.

I informed the NICU staff that I wanted the RT to face serious consequences for her actions and that they needed to make sure I didn't see her as I wasn't going to be responsible for my actions if I did. I literally wanted to choke the shit out of her. I was so angry. I had no problem letting the hospital administration know my feelings either.

Sophia's nurse was amazing and was in tears because she wasn't there to stop the RT from what she had done. I felt horrible for her but at the same time I found comfort knowing she cared that much about Sophia. She literally cared for the babies as though they were her own. She was an angel in my eyes.

As the days and weeks went by, I visited the NICU every day except when I got a cold. For five days I was banned from visiting because I was sick. All I did was cry and call constantly.

During Sophia's stay, she would just overcome one hurdle only for us to learn of another we needed to worry about. Initial brain scans showed Sophia's corpus callosum had not developed and that was a critical part of the brain that allows the left hemisphere to communicate with the right hemisphere. Children born without them are likely to be severely disabled. I knew this because Nichole, my little sister from the Navy, had married and had a child born without one. The fear is real when you know someone who suffers from a debilitating condition.

Sophia also had a PDA or patent ductus arteriosus. PDA is a heart defect found in the days or weeks after a baby is born. All babies are born with this opening between the aorta and the pulmonary artery, but it most often closes on its own right after birth when the baby takes their first breath. It really is a miracle when you think about it. There is medicine they can give a baby up to three times to close the valve. However, if it doesn't close after the third attempt, heart surgery is required. Thankfully, Sophia's closed after the third dosage.

In total Sophia had four blood transfusions because of the blood loss due to constant samples being taken for lab testing. I begged them to take my blood since it was considered a universal blood type, but they said it couldn't be tested in time and it was against protocol. The nurses assured me the blood was safe as they donate it themselves regularly and it's marked just for the NICU babies. I'll never know if that is really true, or if they just told worried parents that to make them feel better.

Meanwhile in the weeks following Sophia's birth, I was struggling to breastfeed. I was setting alarms to pump every few hours and was producing less and less. I was in constant tears from the pain and my boobs were ridiculously big. Then one day when I walked into the NICU, the doctor pulled me to the side. He was concerned about me as I had become a zombie just going through the motions of everything. He asked me why I wasn't getting any rest and I just began to cry. I told him how much pain I was in and the frustration of not being able to keep up with the milk needed for Sophia. The doctor took away my guilt and told me it wasn't meant for everyone and not to worry about the Nipple Nazis. He told me at the rate I was going, I would be the one in the hospital when it was time to take Sophia home. He told me she had gotten all the benefits of the liquid gold and it was okay to quit. I was sad that I wouldn't get that bonding time with her when she was ready to feed on her own, but I knew the doctor was right, I was a mess.

For the next several days as I quit pumping, the pain became excruciating, so I was allowed to take pain pills as the milk dried up. Somewhere someone told me stuffing my bra with cabbage and wrapping my boobs in the leaves would speed up the process. So there I sat, looking like the human version of coleslaw praying my boobs would go back to normal. After about a week, they finally did.

Unlike most first time mothers, my baby shower came after Sophia was born. I spent the day at the NICU visiting her and when I got home, Tina and my mom had invited all of my friends over to throw me a shower. It was a much needed celebration as everyone played silly games and laughed. One girlfriend made the cake and others helped decorate. Between all of my friends, I didn't need a thing when they were done. Everything a first time mother could possibly need was there. I felt truly blessed.

Meanwhile I was worried about my time off from work. As the weeks passed by, I was very close to running out of sick and vacation time. My employer put out a request via email asking if anyone would be willing to donate their own time so that I could continue to stay out of work to focus on Sophia's needs. Between five guys, three SWAT and two K9 officers, three months worth of time was donated. Just enough to get Sophia home after she was released and then start taking her to the countless specialist appointments every day that come with having a preemie.

I made sure to thank each and everyone who supported and helped me during that challenging time. They say it takes a village to raise a child, who would have known I'd need an entire law enforcement agency just to have one.

Little did I know a few years later, a new boss would step in after our great leader retired and our agency would change dramatically. We would no longer be given the opportunity to take care of one another. All he cared about was pinching pennies which would create tremendous hardships for countless officers. Over the next several years, officers were leaving in droves for other, better agencies as we became overworked and underpaid. I will never understand how leaders fail their people. When will they understand the concept of "Take care of your people and they will take care of you"? Many believe the "thin blue line" represents cops looking out for each other no matter what they have done. What it actually represents is the thin line that separates us from protecting the good from evil. No cop wants to cover for a bad one. We understand what it does to our profession and none of us are willing to risk our career for someone else's greed or stupidity. Rather than leaders explain this concept to the media and those who don't understand it, they seem to hurt the very people out there risking their lives all in an effort to further their own careers.

My advice to anyone voting in an election is to forget party lines and politics. Do your own research on each candidate. Ask how the people working for them feel. Find out what the candidates have stood for in the past and what, if any, controversy has followed them. We all have a civic duty to vote for our government leaders and we should do our own real research in an effort to make a smart decision. Lastly, use caution when considering union endorsements, they can also be very political and don't always reflect the feelings of the majority they represent.

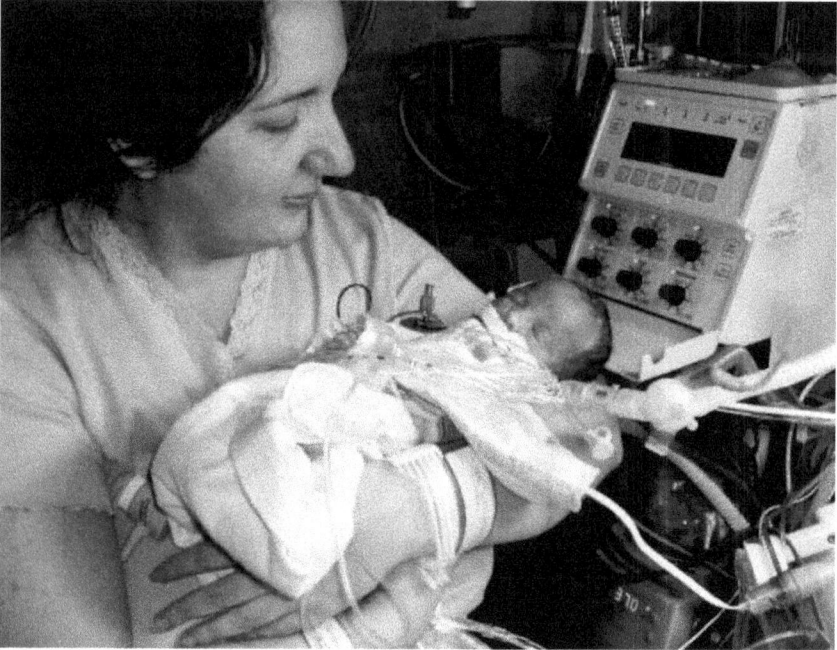

After three days, I finally held my daughter for the first time while her nurse changed her bed sheet. It was fast, but extremely emotional

Seventeen

❦

The Joys & Challenges of Motherhood

Florida, 2005 - 2014

I've often wondered if Sophia's time in the NICU happened for a reason. I mean, ME?? A Mother?? Most of my close friends who had known me forever took secret bets during my pregnancy as to how long the baby would survive. Not because I was a bad person, but because I was REALLY THAT CLUELESS when it came to children. I was the youngest of all my cousins, an only child, never babysat anyone, never changed a diaper and if someone offered to let me hold their baby...I would politely decline for fear of dropping them or hurting them. I had absolutely ZERO maternal instincts.

During Sophia's two and a half months in the NICU, the nurses taught me everything. How to hold her, support her head, put her to sleep, feed her, change her diaper, bathe her, literally everything. I learned that SIDS (Sudden Infant Death Syndrome) was often the result of babies smothered to death because of the type of bedding they slept on or loose items left in their cribs.

They taught me about the dangers of cigarette smoke, even the odor of it on someone's clothing could be a hazard, an issue that would eventually drive a major wedge into the relationship I had with my mom. I learned about viruses like RSV (Respiratory Syncytial Virus) which are common everyday viruses that cause colds in most people but are damned near a death sentence for a premature baby whose lungs haven't fully developed. Sophia would need a monthly vaccination shot for the RSV virus for the first one or two years of her life at over $1000 a pop! It was crazy!

By the time Sophia came home, I felt as prepared as I could possibly be. She had finally reached the weight required to go home, a whopping 5 lbs! She was still having episodes of Apnea and Bradycardia where she would stop breathing and her heart rate would drop, so doctors sent her home with a heart monitor. Literally everywhere we went, we carried this heart monitor about the size of a brief case. My daughter was constantly plugged into a wall to keep the battery charged for when we needed to go somewhere. Turns out every time she bared down to poop, the monitor would go off. Initially this sent me into a frenzy, then it became comical as I learned her schedule and laughed wondering if I was the only mother in the world who knew it was time to change a diaper because of loud alarms going off. Sophia absolutely hated having a dirty diaper, so in addition to the alarms, I got blood curdling screaming to go with it. Funny thing is, for some strange reason, I enjoyed changing those diapers because she never cried for any other reason and the instant I took the diaper off and wiped her clean, she would look at me as though I was her hero and her smile and cooing was her way of saying, "Thank you." Little did I know her mind would change when she became a toddler. She relished running from me with a load swaying side to side from her adorable little butt.

Sophia never cried to sleep, eat or any other reason other than a dirty diaper. She did however have extraordinarily animated facial expressions. She was easy to read based on those alone. Before she learned to speak, her eyebrows would speak a thousand words. I never got tired of holding her and interacting with her. She might have made me feel like a hero, but the fact of the matter was, she was mine. I had become extremely attached to a song by Martina McBride, "In My Daughter's Eyes." I would listen to it everyday when I drove to the NICU and as often as I could once we got home. It perfectly depicted how I felt about her. She was making me a better person because I felt like I needed to set a good example for her.

While my friendship with Zach continued on a platonic level and we spoke on the phone from time to time, I swore men off completely. I didn't want to be one of those single mothers who constantly had a different man coming in and out of their life. I read several articles that said daughters of single mothers will emulate what they see later in life and that was the last thing I wanted for Sophia. I wanted to spare her from the heartaches I had endured.

During my maternity leave and when it was safe to do so, I took Sophia by the office to see some of my coworkers who were asking to see her. During that visit, this one particular woman, the captain's secretary, approached me and explained that she had not put out a birth announcement like the agency typically did for everyone because they weren't sure if the baby would survive. I was hurt and angry beyond words. Was that something she really needed to share with me? Even if there was a question as to Sophia's survival, wouldn't it make sense to put the word out so that people could keep us in their prayers or offer words of encouragement? I was feeling so alone and overwhelmed, this was the most difficult time in my life and God knows, I could have used the moral support. I never forgave that bitch. I hated her with a passion. It seemed as though that was the turning point for me when my career no longer defined who I was. Being a cop was now just my job, not who I was. Being a mother became who I was and I decided I would never let the job dictate otherwise.

Between taking care of Sophia and shuttling her to countless doctor's appointments, I had no idea how much weight I had gained nor did I know how much I weighed when I gave birth. I had been on bed rest and no one was monitoring it. What I did know was that by the time I was preparing to go back to work, I was still wearing my maternity clothes. After all, they were comfortable and I was so busy taking care of Sophia, I hadn't even bothered to look in a mirror or get on a scale.

The truth was, I didn't want to go back to work. I was enjoying being a mom way too much. I worried about my safety when I went back and who would care for Sophia if anything happened to me. My girlfriend from work had her baby about a month after I did. We spent a lot of time on the phone and got together for baby dates as much as we could. She felt the same way I did. We had this brilliant idea of starting our own daycare for first responders and hospital workers who work nights. After hour child care remained a challenge for those who worked in law enforcement, firefighting and hospitals. We had it all figured out, until it was time to do the paperwork. That's when we realized what a challenge it would be to keep up with all the permits and licenses needed. We faced our reality that going back to work was inevitable and we would just have to suck it up.

Returning to work was rough. By this point I had come to realize that going back to be a detective in major cases was a horrible idea and I resigned knowing it was the best thing for both Sophia and me. Major case detectives are married to their jobs and have to be able to drop everything at any given time to go to work, most of which is in the middle of the night. There was no way I could do that with Sophia, nor did I want to.

Soon after returning, I had to get cleared with a physical because I had missed my biannual physical while I was pregnant. There I was met with the same doctor we had all been seeing for years. A tiny little old man who might have weighed 100 pounds soaking wet. The first thing he said to me when he walked in with my chart was, "So tell me, how did you gain so much weight since we last met?" Probably not the nicest thing to say to a woman who just had a baby, so I responded in my typical sarcastic tone, "Cut me some slack...I just had a baby." He congratulated me and asked me when I had given birth and I told him. He then lectured me about how it had been six months since I gave birth and I should have lost the pregnancy weight by then. Needless to say, I wasn't amused, so I replied, "and how many kids have you popped out, Doc?" I truly had let myself go with everything that was on my plate, I just didn't want to face it.

The following Friday I got a phone call and it was the doctor. I'll never forget his words, "Donna, this is Dr. Ponty, I'm calling about your weight." I flipped out and started yelling at him that he didn't need to be calling me at home to nag me. To which he apologized for upsetting me and continued to explain himself. He told me he was concerned about the 80 lbs I had gained so he ran some extra labs on me. He continued to say that I basically had no thyroid activity at all which meant no matter what I ate or did, I would continue to gain weight and struggle to lose it. *Well isn't THAT fan-frigging-fantastic* I thought to myself. Needless to say, now I had to follow up with an endocrinologist and get on thyroid medicine to level myself out so I'd stop gaining weight. It also explained the hair loss and fatigue I had been noticing which I blamed on stress.

I managed to get assigned to patrol where I lived so I was never far from home and could stop by during my shifts to use the restroom, eat and check on Sophia. I was assigned to the evening shift and Mom did a good job putting Sophia on a sleep schedule that coincided with my work schedule so I could sleep in the next morning, but we continued to butt heads over her smoking. No matter what I said or the doctors would tell her, I couldn't get it through her head how harmful it was for Sophia and her premature lungs. Finally I had to put my foot down and tell her that she either stop smoking in the house or find another place to live. I cried as I begged her not to force me to pick between her and my child. She didn't think I would kick her out and I told her not to force me to prove it. She knew me well enough to know I was headstrong and I'd probably do it, even if it killed me. She started smoking outside which was good, but bad in the sense that she spent most of her time outside instead of tending to Sophia when I wasn't there. It drove me crazy but there was nothing I could do. I was stuck between a rock and a hard place when it came to child care. I literally had no one to help. I managed to schedule all of Sophia's doctor's appointments on my days off and never had to worry about getting called into work.

Approximately four months after returning to work, an opportunity came up to work in the schools with kids. I decided this would be a good idea for me to get out of the line of fire at work so I could be around to see Sophia grow up. In addition, I loved the idea of working with kids and mentoring them so that they wouldn't go down the wrong path in life. Quite a big change from the person I was before motherhood.

Once I was selected for the position, I had to wait until the next school year started to transfer since it was summer and school was out. I was training a new officer when a call went out of an 11 year old girl having a seizure. We were really close to the home and got there very quickly. When we went into the home, the mother was hysterical on the phone talking to 911 and her daughter was lying on the living room floor not breathing. My partner and I quickly started CPR and managed to bring her back. Then she crashed again and we started CPR again. We repeated this process three times before the fire department finally got there. I was angry at this point wondering what was taking them so long so the second I saw them walk through the door, I yelled at them to hurry up and help. They have far more training and equipment to help with a situation like this and I was frustrated that my partner and I weren't able to revive her and keep her going.

The paramedics loaded her into the ambulance while continuing CPR. They transported her to the nearest hospital and once they arrived there, the hospital staff took over. The pediatric specialist was on call in the ER that night and I knew him from a stay Sophia had for difficulty breathing one day. I was struggling with my emotions because this time, I was a mother and after everything Sophia had been through and the emotional roller coaster of the NICU, the feeling of loss was much more personal than any time in the past before I became a parent. The doctor told me the little girl had a degenerative brain disorder that was causing her to lose control of her involuntary reflexes. As a result, she had lost the ability to swallow and consequently, was slowly drowning in her own fluids. There was nothing that could be done other than put her on life support, make her comfortable and give her family an opportunity to prepare themselves to say goodbye. My heart broke for that mother. Apparently they had taken their daughter to a number of doctors to figure out what was wrong, but before they could get definitive answers, this happened. It was devastating news.

I excused myself to the staff restroom and called my mom, crying my eyes out. All I wanted to do was go home and hold Sophia. I was a mess and I didn't know how many more of these tragedies I could handle.

Once we cleared that call, it was back to work as usual. At the end of the shift, my sergeant called me into the conference room where the squad was waiting with a big cake. It seemed it was my last day working patrol and I could start my new assignment in the schools soon. While I was glad I was leaving, I was still silently grieving for that little girl and the pain her parents were going through.

Meanwhile, at home, Sophia was getting occupational therapy three times a week. She was having trouble meeting her milestones. We had the sweetest little old lady who reminded me so much of my ornery grandmother. She would get on the floor and work with Sophia. She helped her learn to roll over, push up and sit. Sophia refused to crawl and wouldn't play with anything that required the use of two hands. Eventually she walked at 18 months and that was when OT could stop, she had officially "caught up" with other children her age. It was the best of times. She was such a happy baby with a laugh that was absolutely contagious. She loved balls or anything that was round and bounced or rolled. She would belly laugh so hard she would have tears in her eyes. Once she laughed so hard during play that she cried later from being sore.

Sophia's appetite continued to be a challenge. She just wasn't a big fan of eating. She was, however, a big fan of sleeping. Every time we would give her a bottle, her little hands would go straight to the top of her head and she would massage her bald little head until she fell asleep.

It was such a challenge to keep her awake to finish it, but we didn't mind, she was too cute for words.

In 2008, after not speaking with my father for over a decade, I received a call from the North Carolina State Mental Institution. My father had been living there for several years after the courts placed him there. It seemed he had spent every dime my mother left him, sold his home and completely lost his mind. He walked into his bank one day with a gun demanding to know who stole all his money. He was arrested and deemed incompetent by the court and placed in the hospital.

After a few years, his years of smoking and his age caught up to him and he was dying of cancer. The doctor on the phone was incredibly nice and apologetic for calling because he knew my father and I were estranged. The doctor told me my father was dying and was in his last hours. He explained my father couldn't speak and hadn't been able to speak for some time, however, he could understand what someone was saying to him. The doctor continued to explain to me he felt my father was fighting death because he was afraid. He kindly asked me if I would speak to my father in an effort to calm him so he could die in peace.

It was an incredibly unexpected phone call. I had been so consumed with becoming a mother and the joy Sophia was bringing into my life that I wasn't prepared for this decision and I didn't exactly have time to decide. I asked the doctor for a few minutes to think about what I wanted to do and what I could say. I needed to process what was happening. I spoke to my mom who was incredibly supportive and felt strongly I should do what the doctor asked. She feared I would regret not making peace with my father before he died.

I called the doctor back and told him I would do as he asked. He put the phone to my father's ear and I told him who I was. For the first time in over a decade I heard my father's voice even though he could not form words. I wasn't prepared to tell him I forgave him or even that I loved him. I had closed that chapter of my life and had no intentions of ever opening it again. Yet there I was, trying to form words to a man whom I'd wished dead several years prior. To say it was uncomfortable would be an understatement, but I did it. Why? Simply because it was the right thing to do. It was incredibly hard and I knew I didn't mean what I was saying, but no one should have to die that way, alone and feeling unloved, so I did it.

My father died 45 minutes later and his cremated remains were mailed to me. I carried them back to North Carolina and had them placed in a special area he loved. I have since found forgiveness for him and I do pray he is in a better place where he no longer feels anger and sadness.

Sophia was such an incredible blessing who brought joy to us on a daily basis. Because of her, I was able to navigate through that confusing and difficult time easier than I would have had it not been for her. One of my favorite memories of her was when she wouldn't go anywhere without her little pink princess purse hooked on her arm. She took it EVERYWHERE SHE WENT! Mom and I would laugh thinking she was never going to give that thing up.

Around the time she was approaching two, I started to notice how she would run a little sideways. I kept mentioning it to her pediatrician who kept telling me to stop being paranoid and just enjoy her. Mom noticed it too and my gut told me something wasn't right. After six months of bringing it up to the pediatrician and explaining how Sophia wouldn't use her right hand at all, for anything, I decided to get another opinion. I took Sophia to a place we had gone that was responsible for tracking her progress through her milestones. They had me bring her in for an assessment and referred me to a neurologist. After only 15 minutes with a pediatric neurologist, Sophia was diagnosed with cerebral palsy. I instantly began to cry because I remembered the twins I had escorted through the theme park a few years back. In my mind, that was what CP looked like. The doctor quickly calmed me down and told me CP came in a variety of shapes and sizes. She told me on a scale of 1 to 10, 10 being the worst case scenario, Sophia was a 1 and that it would never get any worse. She told me with hard work and exercise, Sophia could overcome the symptoms of CP and by the time she was in high school, no one would notice. She prescribed physical and occupational therapy for her and we quickly made arrangements like we had before for home visits.

I called Sophia's pediatrician and made an appointment with her. I informed her that one of Sophia's neonatologists gave me great advice when she was born. He told me a good doctor will always listen to a mother's concerns because mothers have a keen instinct when something isn't right with their child. He told me to always trust my gut feeling and if a doctor didn't listen to me, to find another doctor who would. I informed her of Sophia's diagnosis and told her she wouldn't be seeing us again. Then I walked out and never went back. A few years later I heard she left medicine and I never found out quite why, but I have my suspicions.

Over the next couple years, Sophia continued to be the best thing that ever happened to me. Work no longer defined who I was, it was just a job that paid bills. Working in the schools was an eye opening experience for me as an adult because I never attended public school until I was in high school. Never in a million years did I ever think I'd be dealing with lesbian love triangles with kids in the sixth grade! I was not trained nor equipped to handle those situations so I quickly directed the kids to their guidance counselors.

It didn't take long before I realized why my parents sent me to Catholic school. I became increasingly concerned with the way public schools taught children and handled situations that were most certainly the job of their parents. With that said, parents also seemed to send their children off to school with the expectation teachers would raise their kids so they wouldn't have to. It was really sad for many, but it helped me learn from others' mistakes when it came to parenting.

When Sophia got diagnosed with CP, I talked to one of my guidance counselors about it. She warned me public schools would put a label on her and pigeon hole her into a classroom with other kids who had all kinds of different disabilities. Schools would do this because they get extra funding for every child with special needs. I was concerned because I knew Sophia was smart, she just had some physical limitations. I was advised to look into a state scholarship program for kids with disabilities so the parents could send them to the school of their choice. I started learning about that program right away and found out that I could, in fact, send Sophia to Catholic School like I attended where she would be put in a classroom like any other normal child. As a single mother on a cop's income, Catholic School wasn't even an option financially, but with this scholarship program it was the answer to my prayers.

Over the next year, Sophia continued to slowly improve and when she turned 3, she was entitled to start Pre-K early to get a head start in preparation for Kindergarten. Between doctors appointments and everything else, I went back to the road because the schedule allowed me to be more flexible during the week. I returned to the same area I was in before so I could stay close to home and that made things a lot easier.

One day, I got a phone call from Sophia. She was four years old and had been able to recite my phone number and our address since she was 2 1/2. Sophia told me her grandma was being mean to her and wouldn't fix her something to eat. I knew that didn't sound right, so I went by the house to see what was going on. There I found my mother acting strange. She was disheveled and angry. She wasn't making sense. Sophia had asked her to make her a peanut butter and jelly sandwich and my mother smeared peanut butter on a plate with no bread. Then she started cursing about Sophia being ungrateful. Initially I thought my mom was having a stroke, but the symptoms didn't add up. Then I felt her and she was burning up with fever. I called for an ambulance since Mom was refusing to listen to me. They came and I told them she wasn't right in the head and insisted they take her to the hospital.

As it turned out, Mom was suffering from a really bad urinary tract infection, something fairly common with post menopausal women. The nurse at the ER told me that when a woman gets a UTI after menopause, it often affects her brain and cognitive skills. Mom was in the hospital for a week as they fought the infection. Had we caught this any later, she likely would have gotten sepsis which would have been life threatening.

I was so grateful Sophia had the sense to call me and knew how. We all agreed she got credit for picking up on "something wasn't right with grandma." As time passed, it became more and more obvious Sophia had an interest in medicine. Her favorite game to play was doctor and we all got regular checkups. She loved going to the doctor and would ask them to explain all the diagrams in the room. She didn't like it when they explained things in a simple way. She wanted to know what all the big words were and she wanted to truly understand how the human body worked. By first grade she could explain the respiratory system and pulmonary system to you and get it right. She loved everything about medicine and the human body and never got tired of learning or explaining.

One of my favorite memories of her at this age was when she had a white board and was using it to draw an X-ray of a sick patient. As she drew with her green marker and explained the hazards of cigarette smoke and the cure to every ailment was green ice cream, she began to look for the cap to the marker. She looked and looked and couldn't find it, that's when she said, "Where's that cap? Where'd I put that damn cap?!" With that I just gasped. I didn't even yell at her, but the shocked look on my face clearly led her to know she said something wrong. Without hesitation, she had instant regret and began to cry apologizing profusely for saying a bad word. I tried so hard not to laugh as I explained to her that was a word only adults could use.

Another time she was playing in her room when she found a polished rock I had gotten her at a theme park. For some strange reason, she thought it would be a good idea to put it in her mouth. Polished rock + saliva = slippery trip to the tummy. Sophia came out to her grandma with fear in her eyes pointing to her mouth. My mom eventually figured out she had swallowed something but couldn't figure out what because Sophia was panicking. Mom called me and the first thing she said to me was, "Now don't panic" to which I replied, "too late, I'm on my way, what happened??" I wasn't far from the house (not a coincidence) and when I walked through the door, Sophia was in tears because she was afraid of what the rock was going to do in her tummy. The second I figured out what she had swallowed, I knew we had to head for the hospital because it was a fairly big rock and I was shocked, but grateful, she didn't choke on it. Once we got to the hospital, they told me, "Oh, don't worry, kids swallow stuff all the time. If it's small enough to swallow, it's small enough to pass." I told them I didn't think so, I knew how big that rock was. They told me "all moms say that." They took her to get an x-ray and then we waited for the doctor to come. As soon as he put the film up, he saw the rock lodged at the bottom of her stomach stuck in the entry to the small intestine. He looked at me, took a deep breath and said, "Uh, yeah...we need to go get that out." He booked an operating room and we waited. Fortunately they didn't have to do any cutting on Sophia, just put her under and went fishing down her esophagus. They took a video of the scope going down and the little net they used to scoop up the rock. Sophia LOVED talking about those pictures for years. She showed them to EVERYONE! The doctor gave us back the rock in a specimen cup. Later a friend had that rock turned into a charm and we call it "Sophia's lucky rock" because she was lucky it didn't kill her.

By the time Sophia was in third grade, the emergency room staff at the children's hospital knew us on a first name basis. They would say, "what did she do now?" To say my daughter was a little accident prone would be an understatement and we could probably write a book, just about those adventures. With that said, I will say our trips did begin to slow down once I caught her exaggerating her pain and trauma JUST so we could go to the ER. Turns out those were fun field trips to her and we had to have a long talk about how they weren't free. Over time she became less accident prone and more understanding of how medical bills work. She also started rendering aid not only to herself, but Grandma every time she needed it.

Needless to say, she is a bright teenager now and has decided she wants to be a pediatric neurologist. She continues to be fascinated with the human body and has won two science contests at school for her research on the brain.

God truly Blessed me with this child and I genuinely don't know where I would be without her. We have our struggles like most mothers and teenage daughters do, but when I look at the struggles of other parents, I know I'm really lucky to have a kid with such a strong moral compass and genuinely good heart.

Eighteen

❦

Taking a Leap of Faith

Even though motherhood suited me and I had a good career, my life experiences somehow kept me from feeling true happiness. I had no hope of falling in love again or any desire sexually or otherwise to date or meet a man. In addition to a stressful and demanding job, my plate was full taking care of Sophia and my mom, whose health continued to deteriorate.

Our neighborhood had begun to take a downward spiral due to crime and I wasn't comfortable letting Sophia out to play thanks to drug dealers down our street. You would think the marked patrol car in my driveway would have been a deterrent, but criminals had become bold and just didn't care.

Meanwhile I couldn't go anywhere without running into someone I had encountered on the job. It felt like I was at work 24/7. Strangers would knock on my door at all hours of the day and night to ask for help instead of calling 911 for their domestic disputes or problems they needed to report. Once I had a guy who got high at a college party and began to hallucinate so he ran to my house, where the patrol car was parked, and started screaming, "I'm going to fucking kill you" as he was trying to kick in my front door. There I stood with Sophia in one hand

and my gun in the other while cradling the phone in my ear talking to 911. I couldn't count how many times I told them, if they don't get there before he kicks this door in, I'm going to shoot him. Thankfully the cavalry showed up and the lunatic got to live another day.

Then there was the day I was at the children's hospital with Sophia for a doctor's appointment when one of my students walked in with her father. Just the previous day I had interviewed her in a child abuse case and he was the suspect. Everything would have been fine until the little girl recognized me and called me by my title. "Hi Officer Donna!" she said and I immediately saw the anger in his eyes. He was tatted up both arms, neck and face and some of them were clearly gang tattoos. I immediately stood up and walked to the reception desk and asked for a private room for our safety. That was it, it was time to move, things had gone too far.

Over the course of seven months, I had three contracts on three different houses and none of them went through thanks to the economy and housing market crisis Clinton created before he left office. The housing bubble had burst and homes were being foreclosed on and sold as short sales at epidemic rates. I was trying to use my VA loan, but the seller's lenders wouldn't budge or compromise with the VA's requirements, so it made things really frustrating.

I gave up looking for a house for a while due to the stress it had created and then one weekend, while visiting some friends in my home town on the coast, about 45 minutes away, I saw a house for sale that was exactly what I was looking for. I called my realtor and told her about it. We made an offer and 28 days later we were moving in. I put my house in the city up for rent in the hopes the market would improve so I could sell it when I could break even.

The new house was everything I could ever want in a home, plenty of room, great neighborhood with wonderful people, peaceful and best of all, it had a pool. Even more special was that Sophia was going to attend the same school I did as a child. It all seemed meant to be.

Sophia was growing up and absolutely loved going to theme parks to ride roller coasters. I had tried every diet gimmick, plan, and pill on the market and all my weight did was take one step forward and two steps back. I was tired all the time, my body hurt, I was afraid I wasn't going to be able to ride the rides with her much longer and I had just given up. I dreaded getting on the scales because every time I did, the number just got bigger. I was 5'6 and I weighed 287 pounds. I tried my best to pretend I was happy and I was in complete denial about how big I really was.

Then one day I was on a social media page where I saw a former supervisor who had struggled with his weight for years. We used to complain about how other people didn't understand the struggle some of us had with our metabolisms. There he was, in his uniform, announcing his retirement. I hadn't seen him in about three years. I almost didn't recognize him. I quickly called him and said, "First congrats on retirement! But you have GOT to tell me how you did it!" He laughed, thanked me and confirmed I was referring to his weight loss. I simply said, "Yes! You clearly found a secret and I need to know what it is." He replied back with four words I never expected to hear, "I went to Mexico." I'm sure there was a pause as I processed what he said and then I replied, "Aaaaand you got Montezuma's Revenge??" He laughed and said, "No," and went on to explain how he went to Mexico to get weight loss surgery because our insurance wouldn't cover it. I told him he was out of his mind and he was lucky he didn't go missing as they sold his organs on the black market. He kept laughing and said he understood where I was coming from because that is exactly what he thought when he heard about other officers going there. Then he commenced to give up names. I was in shock, here were all these people I knew that were going around saying they "just ate less and exercised." I was angry at first thinking *what a crock of shit* that was for them to lie about how they lost weight, then I realized it was a personal matter and I suppose I could see why some wanted to keep it private. Brad on the other hand knew me and understood my struggles, so he was honest. He gave me information on a point of contact, a website and was willing to answer all my questions. He had really given me something to think about.

After six months of doing my research on the private weight loss hospital located hundreds of feet from the California border, I couldn't find a single complaint from anyone. The vast majority of their patients were from the U.S. and Canada and all anyone would say was what an incredible experience it was and how wonderful the patient care was. I went out of my way looking for negative reviews and complaints and couldn't find any on this specific hospital. So, I took out a personal loan to borrow the money, which was less than a third of what U.S. Hospitals were charging for less patient care, and I went for it. In the U.S. the surgery is considered outpatient and there are no tests for follow up to catch any complications that may come. In Mexico, they do the surgery and then keep you for four days to run tests to make sure everything is safe before they discharge you. I was truly impressed and I had come to realize that I was too young to be living the lifestyle I was and I had a daughter who needed me to stick around in the years to come.

On December 1, 2015, I boarded a plane for San Diego where a driver would pick me up and drive me two hours to the Mexican desert where the hospital was located. Some pre-op tests were done and then I was taken to a five star hotel where I got to eat my "last meal" for a while. The hotel was beautiful and the steakhouse was fancy, but I was so nervous about the surgery, I had no appetite and I just couldn't eat. I was given a sedative to help me sleep, because apparently fear is normal in these circumstances and they didn't want anyone fleeing the hotel in the middle of the night into the Mexican desert.

The next morning I was picked up and taken to the hospital for surgery. Everything went smoothly and the first day I mostly slept. I tried my best to talk to Sophia, but I don't remember it. Sophia said I just kept falling asleep on the phone and then the nurse would hang it up for me. The next day I woke up and could see the three tiny incisions on my belly as the entire procedure is done laparoscopically. They had removed 80% of my stomach and converted into what they call a vertical gastric sleeve. The procedure is designed so a person can only eat a fraction of what they could before. To be safe, the first two weeks the patient is on clear liquids only while the stomach heals, then soft liquids like yogurt for the next couple weeks, finally soft solids for a while and then regular solids. The patient has to work very hard to take in the necessary protein everyday, which when you have a 3-4 ounce stomach, is a challenge.

After my four days were up in Mexico, my driver drove me back to San Diego where I boarded a plan for home. I couldn't wait to get there because I missed Sophia so much. Once I got home, I had no appetite at all. Staying hydrated and getting protein was a challenge, but I kept really good records daily of what I took in, sometimes I met my goal, sometimes I fell short. It's amazing how much the stomach can hold and you don't even realize it until someone reduces it down to size. I learned I couldn't drink when I ate or the liquids alone would take up space. I made sure not to drink 45 minutes before or after eating so the food in my stomach wouldn't swell and cause pain. My body went through some crazy changes over the course of the next several months after the weight came off.

It turns out that estrogen is stored in fat cells, so when I was a couple months post-op, and the weight was coming off quickly, I would get estrogen surges. I really wished someone had told me this was going to happen so I would have been psychologically prepared for the consequences of it. One day I was sitting in an incredibly dry boring class at work when suddenly my mind went straight to the gutter. I hadn't been horny in years and there I sat trying to focus on computer technology in law enforcement when I couldn't get my mind off of hot steamy sex. It was beyond distracting and frustrating. A few hours later, I was sitting in the parking lot of the grocery store sobbing and crying and I didn't know why. I was a total mess and I thought I was losing my mind.

After going through similar episodes of this over the next few weeks I mentioned it to a female coworker who had the surgery done before me. She trusted me enough to tell me after I told her I was going for it. She laughed and said, "oh yeah, I should have warned you about that." It didn't necessarily stop the hormone surges, but at least I knew I wasn't losing my mind when they happened.

By the following summer I had lost 100 lbs and I felt 20 years younger. My joint pain was gone, I had energy and I felt good about myself for the first time in years. Shopping for clothes suddenly became fun and I never got tired of the compliments I got on a regular basis. People who hadn't seen me in a while truly didn't recognize me.

Then one night I was saying goodnight to Sophia and reminding her to say her prayers when she said, "You know what I pray for Mommy?" To which I replied with a laugh, "Oh here we go, what is it this week? Another dog? horse? cat?" She said, "No, I pray for you to find someone to marry so when I'm all grown up, you won't be alone." I choked back the tears as I kissed her goodnight and then I went to my room and cried my eyes out. I wasn't stupid, I knew why she wanted me to date, she wanted a dad. Her father still remained absent in her life and it was so obvious she didn't understand why everyone had a dad and she didn't. Plus she had a point, someday she was going to grow up and be gone...and then I would be alone. She wasn't the first person to say this to me, but she was the first person to get through to me. Did I really want to be alone?

Even though Zach and I remained close friends and talked regularly, it was becoming more and more clear he was never going to leave his wife and I was a fool to have wasted so many years of my life waiting for him. I was 46 years old and still single.

I made an attempt at meeting someone on a dating site, but it was beyond frustrating. Most guys cancelled at the last minute or never had time to meet. Personally, I think a lot of them were married and just curious about what was out there while looking for attention they couldn't get at home.

I met a couple nice guys but there just wasn't any spark or something just didn't seem right about them, so I wouldn't go on a second date, why waste my time or theirs?

Then I got a phone call from a good friend. He was planning a surprise party for his wife's 40th birthday and needed help decorating the house for her. I agreed to help and was excited to spend time with them because I hadn't seen them in a while.

When I got to their house, I started decorating and preparing things when some of his friends showed up to help with heavy lifting. The second I saw Duane walk through the door, I was smitten. He was tall, good looking and had that manliness about him that I was attracted to. As the evening progressed, I quickly learned I had to be really careful drinking as I was an even bigger lightweight than ever before. I paced myself through the night so I wouldn't ruin it.

Duane and I spent most of the night talking and we really hit it off. We were both single parents, I had my daughter full time and he had a daughter a couple years older who he also had full-time. He had been raising her with the help of his mom. We had a lot in common and I was incredibly attracted to him. I don't recall how it came up, but we started discussing ages. As it turned out, I was much older than he thought I was and he was much younger than I thought he was, 13 years younger to be exact.

Then something happened that caught me by surprise, he gently put his hand on the small of my back and my knees buckled. I quickly realized how much I had missed the touch of a man and it was hitting me like a freight train. The next thing I knew, we were making out like teenagers and he was doing his very best to go for a home run. All I remember was repeatedly telling him "No, I can't." While I was feeling the alcohol and hormones, I was also keenly aware that it had been 11 years since I had been intimate with anyone, the last of whom was the love of my life. There was no way I was going to do something with a guy I just met. I knew I would regret it. Then suddenly it was like my buzz just up and left. I felt sick to my stomach and I started to have a full blown panic attack. I jumped up, grabbed my purse and shoes and literally ran out the door. It was 5 a.m. and I was an idiot for trying to drive, but I felt like I just had to escape.

When I finally got home, I went straight to my room and just cried myself to sleep. What was wrong with me? I felt so pathetic and guilty. I felt like I had cheated on Zach which was utterly ridiculous, but that was how I felt. I called a few friends that day for advice, and they all tried to talk sense into me, but my heart just wouldn't listen.

I called our mutual friend and told him not to give Duane my phone number. That I almost made a big mistake and I needed time to figure myself out. He lectured me like a big brother and called me out, but it didn't do any good.

A few months later a hurricane came through and my house took a pretty good hit. I needed a new roof, pool enclosure repairs, fencing replaced, typical outdoor repairs. I remembered Duane had a background in construction and had offered to help me anytime I needed it. I was desperate for help and needed it soon.

I called Chad and asked him for Duane's number. I called Duane and apologized for the way I acted the last time he saw me. He quickly apologized for whatever he did that made me run. I assured him it wasn't him, it was me and I got scared and freaked out. I told him why I was so scared and how I had been living the life of a nun. Then I told him about the help I needed and he agreed to help me. I offered to pay him, but he refused, so I offered to cook dinner for him while he was here fixing my screens and fence.

That weekend he came over and the second I saw him, I knew I was still attracted to him. Apparently it wasn't the alcohol talking that night. I remember going to the hardware store and feeling a force that just drew me to him as we walked down the aisles.

When we got back to the house, I called my best friend and told her of my dilemma. I snuck a picture of him and sent it to her. She told me I had waited long enough and to just go for it. This was crazy, there was no way I was doing THAT with Sophia in the house, but I did want to spend more time with him.

I walked outside and asked him if he would like to hang out after dinner so we could just talk, then I kissed him, out of nowhere. I'll never forget the look of surprise on his face. I told him I just wanted to get to know him better and he accepted.

Later that night, we spent hours talking. He told me about the challenges he had faced growing up with alcoholic parents and the troubles his sister had created because she was a bipolar drug addict. He told me of the problems he's had with ex-girlfriends and all the drama he'd been through.

I told him I really liked him and that I could tell he had a good heart. I could relate coming from a troubled childhood and that sometimes we all needed a little help getting our shit together. I told him I could see him owning his own business because of how hard he worked and how talented he was. He had no business sense, but that was okay, he had the talent and work ethic to succeed.

Over the next few days, he texted me regularly and we talked every night. I felt like a teenager and it felt good. He was doing a great job sweeping me off my feet.

Meanwhile, the deadline Zach had given me was approaching. He had set a deadline of when he would leave his wife and it was clear he was struggling with pulling the plug. As Duane and I grew closer, I had to make a decision on whether to take the relationship with Duane to the next level or not, so I put Zach on the spot and made him make a decision. It wasn't easy, we both cried and he finally admitted he just couldn't bear taking the chance of his daughters hating him and disappointing his entire family. I was hurt, but deep down I knew he wouldn't, even if I believed he wanted to. I knew the divorce would financially kill him. Zach had put his wife all the way through college and she had a Master's Degree, but chose to never work. It pissed me off to no end. For all those years he worked his ass off providing, now he was retired and still working his ass off. He had even taken a second job to help support his granddaughter and daughter through school and she still wasn't working. She had him right where she wanted him, if he left her she would take half of everything and he ran the risk of losing his kids. I don't know many men that would take that chance.

I told Zach I had met someone and now that I knew where he stood, I was going to give this new guy a chance. He wished me well and reaffirmed he only wanted to see me happy.

Before I knew what was happening, Duane was coming to the house on a daily basis and we were talking about him and his daughter moving in. His landlord was selling the house he rented so he needed to find another place to live. We figured since he was at my house everyday, it made financial sense to share the bills. I met his daughter, Emma and we introduced our girls who immediately connected. They were each so excited to have a sister.

Duane moved in and the entire time, I had anxiety over how fast things were moving. It didn't take long before he started to show his dark side. It seemed he didn't actually make as much money as he said he did, in fact it was a fraction of what he said. He had serious spending problems and while all of my bills quadrupled, his income barely covered a fourth of them. I was suddenly the mother of a teenager I barely knew and it was exhausting because I didn't feel it was my place to make many of the decisions asked of me when it involved her. I was suddenly the bad guy all of the time being forced to say no to things she wanted to do which caused a lot of tension. Duane never took it out on me, I was the sugar momma and his meal ticket it seemed, so he would just drink and completely overreact with his daughter out of fear she would cause me to break up with him. I was quickly learning I had made a huge mistake, but it was too late, they had moved in and their dysfunction was quickly disrupting our peaceful life.

As tensions grew, Duane started drinking even more. As it turned out, he wasn't much different than his father and he was one hell of a mean drunk. Despite that, I kept trying to "fix" him because he reminded me with every sober apology how much he needed my help. He just kept asking me to be patient and would occasionally mention marriage to which I was not even open to discuss. I might have been in love, but I wasn't stupid.

Then I started catching on to the lies. The one that should have sent me running for the hills was the lie about being wounded in Afghanistan when he was in the Army. I started doing some digging and figured out he had never even served in any branch of the military. Now stolen valor is a huge pet peeve of mine, but I was thinking with my heart and not my head by this point. I had fallen into the very trap someone warned my mom about years before. Girls who grow up in abusive homes are likely to find themselves in abusive relationships. I knew this and thought I was smart enough to avoid it, but there I was, deep in over my head.

As Christmas approached, I was buying things for the girls. Emma, his daughter, desperately needed a computer for school, so I bought her one. Both girls got everything they wanted. Duane went out and bought me a diamond bracelet, which I loved, but despite me asking him to wait until Christmas to give it to me, he insisted I open it early because he wanted to know if I like it so he could go back and get something else for Christmas Day.

Two days before Christmas, Duane's entire demeanor changed. He was agitated, mean and distant. I figured out he didn't have the money to buy me what he wanted so I just kept telling him how much I loved the bracelet and that was enough.

Christmas Eve came and Sophia and I always attended Christmas Eve Mass. Duane took Emma out for some last minute shopping with my truck and hadn't returned. By this point I was pretty angry with his attitude and he was giving me a hard time about my Christmas traditions. Neither he nor Emma were "church people" and they mocked Sophia and I for our Faith. I kept hoping they would see we had it good because of our Faith, but they refused.

Finally they got home just in time and Sophia and I left for church. I remember sitting in Mass crying and trying to hide it because it was painfully clear that I was in a toxic relationship and it wasn't healthy. I started trying to figure out how to end it as soon as Christmas passed.

That night we let the girls open a present and Duane just sat as if he wanted to kill someone. When we went to bed, I was left to assemble my daughter's toy by myself. He refused to help with anything.

Christmas morning, we got up and everyone opened their presents. Duane's birthday was the next day and I was saving his toolbox for that since he was so certain I bought it for Christmas. Surprising him was always a challenge because nothing got by him.

Everything seemed to be going better when he left to take his daughter to the city to spend time with her grandparents. We waited all day for him to return and dinner was getting cold and dry. Finally he pulled into the driveway and when I went out to see what took him so long, I knew right away, he was drunk. As soon as I called him out on it, he lost his mind. He started yelling at me so I turned to walk in the house when he started throwing Christmas presents he brought from his mother. A few of them hit me before I could make it in the door. I sent my mom and Sophia to her room and told Sophia to listen to music on her new headphones. I gave my mom the phone and said to listen up for me if I yelled for help. I told her to call 911 if I did that.

Duane stormed in the house continuing to yell obscenities and threats at me. I grabbed my phone and called our friend Chad to come get him because he was drunk and out of control. Duane refused to go anywhere saying it was "his house" and the only way he was leaving was if he was going to jail. With that he raised his shirt and I saw his hunting knife on him. I had backed up into my nightstand where I kept my gun, in case I needed it. I was hysterical, crying and begging him to leave. I kept reminding him it was Christmas and please not to ruin it, but it was too late. Chad told me if I was afraid, to go ahead and call the cops, so I did.

I don't think most people will understand how difficult it is for a cop to call other cops for a domestic matter. It's humiliating and embarrassing, it was the most difficult phone call I had ever made. I didn't want him to go to jail, I wanted him to get help. I told the police that when they arrived but he managed to pull himself together by the time they arrived and made me look like the one who was crazy. They told me unless I could give them a reason to arrest him (I could, but I didn't want to), then there was nothing they could do. I disagreed with that, but my experience meant nothing to them, which frustrated me that much more. They did convince him it would be best to leave however, so he called a friend and they left.

Mom and Sophia ate dinner, I was too upset to eat. Duane had officially ruined Christmas, our favorite holiday and the happiest day of the year. Even my father had never fucked up that bad.

I cried for the rest of the night and Duane just kept calling me apologizing and crying as he asked for forgiveness. I rented a U-haul truck the next morning, friends came over and we packed up all their stuff. I drove the truck to his mother's house where friends met me to help unload it. It was awful. It was his birthday, but I just couldn't live like that.

A week later, like a dumbass, I took him back because he convinced me he would quit drinking and get counseling. I should have known better, but friends convinced me all couples go through stuff and if you really love someone, you have to put in the work.

Things got better for a couple months and then it started again. The drinking, the craziness, the bipolar roller coaster. Finally, one day I was done and he knew it. I got home with my daughter and he stormed out of the house with a pipe in his hand threatening to kill my 13 year old German shepherd, whom I loved tremendously.

That was it! I sent Sophia to a neighbor's house, I went inside and commenced to tell him off. As I went into my bathroom, he started demanding I give him his guns. I had locked them up months before because of his irrational behavior when he drank. I refused to give them to him and he threatened to call the police on me for hitting him. Right then and there, I knew he had to go. Now he was making shit up and jeopardizing a career I had worked incredibly hard at for 17 years. I grabbed my purse and the dogs and I headed for the truck. He was literally chasing me down the street begging me to stop. He knew he had gone too far.

I called a friend and asked if Sophia and I could stay for the night while I figured things out. I called into work the next day and went to the courthouse for a restraining order. It was humiliating to write in a statement some of the things he had done to me all the while I asked myself how I let it happen.

When I got home, he was gone. He had taken his clothes and found a place for his daughter, leaving the bulk of his stuff at the house. It would take me nine months and an attorney to finally get his belongings out of my house.

Meanwhile, after the dust settled, we agreed to try and be friends. I told him I just couldn't live with him until I knew he had his shit together. He agreed to work on himself in the hopes a year down the road, we would be in a better place. I was trying to keep things amicable because he was driving a truck I was dumb enough to put in my name and now owed me over $7000 in bills I had paid for him. In addition, he had nowhere to go with his daughter, so I let him live in my RV on a friend's property while he remodeled the home there.

Despite all he had put me through, I still loved him and I still wanted to see him become a better man. He had an incredible work ethic and if he could just get his head and heart together, I had no doubt he could become successful one day. Sadly, though, he lacked the self confidence to see that in himself and after doing my research, I learned I was not the first woman he had done this to and I wouldn't be the last.

Nineteen

❧

A Perfect Storm

Duane and I were getting along as friends and most of the drama had ended for the most part. I was still trying to collect the money he owed me and we were working on a way to get the truck transferred to him.

Max, my beloved German Shepherd passed away days before my birthday. It was one of the toughest decisions I'd ever made, but I knew it was time and he was suffering. He no longer enjoyed the things he once did and had stopped getting off his couch to greet me when I got home. When I would take him on walks he would begin to limp and it was heartbreaking. I didn't want to make the decision, I so badly wanted him to pass peacefully on his own, he had been such an amazing dog and we loved him so much. I rescued him from a vet office where he had been kenneled for five years after being abandoned by his owner when he was two. The veterinarian kept Max for the purpose of using him as a blood donor dog and would use his blood to donate to other dogs who had been injured. I had no problem with a dog donating blood for the purpose of saving others. What I did have a problem with was kenneling them for five years during the prime years of their life. German Shepherds are a special breed. They need a human to attach to and bond with. Max was terrified of storms and we wondered if it was from being alone after hours in the clinic during storms. We would never know, but what we did know was that we were going to spoil him rotten in his golden years, which we did.

The pain of putting him down was horrific and I hated that Sophia had to experience the loss like I did when I was a kid. Nonetheless, it was time and there was no way to avoid it. I wish someone had prepared me that not all dogs go peacefully. At the last second, as the vet administered the last drug, he fought it and tried to stand up before collapsing in my lap. It wrecked me and I just laid on the floor wailing. Sophia sat in the waiting room because she didn't want to be there. Later she told me she had to put her fingers in her ears to keep from hearing me cry. In hindsight, maybe I shouldn't have taken her, but I didn't want her to be upset with me like I was at my parents when they left me behind the day they put Barney down.

I tried my best to focus on the fact that he was no longer in pain and running through emerald green pastures, but the loss was still there. I immediately took his couch to the curb because it was too hard to see it sitting empty. I rearranged the furniture hoping it would help, but it didn't.

Sophia and I tried to stay busy that week as it was the last week of school. Mom was in the hospital and had been for a week. She had been in and out of the Intensive Care Unit (ICU) and doctors were having a hard time getting to the bottom of what was wrong. Some thought it was her medications, others thought it was oxygen deprivation from arterial damage, it seemed like a mystery as they tried to figure out.

Friday came, it was June 1st, 2017, a day I would never forget. School had ended the day before and several of us were assigned to work patrol that summer to help deter crime that spikes when school is out. Everything was business as usual when a call went out for a drowning. I was just a few blocks away as were two other officers. They found the house a little bit quicker than I did as they knew the area better than I since they worked patrol full-time. As I pulled up to the house, I ran in and found one officer wrestling on the floor with a woman who had a knife in her hand. They were speaking Spanish, she was crying hysterically and as I stood there for a second trying to decide how to help him, he managed to get the knife away from her. It all was happening so quickly. He told me to go in the back to help the other officer. I ran out the back door and there I saw the other officer performing CPR on a two year old boy. The caller was a neighbor who had jumped the fence to help after hearing screaming coming from the home. The caller was standing there with the officer as he tried to revive the child. The child was blue and had foam coming from his mouth. It's a sight no parent should ever have to see. The father, also hysterically crying, was trying to get to the boy but the officer had to keep pushing him away so he could help. I pulled the father away and made him turn his back to the boy because watching CPR performed correctly on someone is a difficult thing to do and it's incomprehensible on a child. I can't even imagine how devastating it would be to witness it on your own child.

I did my best to keep the father focused on me as I asked him questions about what happened. He spoke just enough English that I could gather most of what he was saying. We could see a toy at the bottom of the pool that we figured the child either dropped or was trying to get to.

The father explained to me that he and his wife worked nights and the grandmother was watching the grandchildren while they slept. The boy was two and the baby girl was about 10 months old. She was inside in her crib and the boy was free to roam the house. No one knew how long he was out of grandma's sight and she didn't speak any English. What we did know is that the father had done everything he could to prevent a tragedy like this from happening. We could see the child lock he had installed near the top of the door leading to the pool, but the little boy had built a ladder with his toys and climbed up the door to unlatch the lock. Grandma never even knew he was gone until she went looking for him.

I didn't respond to the hospital, but I heard it was horrific. Someone had allowed the family into the trauma room while the hospital staff was trying to revive the child. Officers had to physically drag the family out of the trauma room just so the doctors and nurses could do everything they could to save the little boy. Unfortunately, they couldn't.

Meanwhile I spent most of the day at the house securing the scene for detectives to investigate. All I could think about was the first drowning and what it was like at the hospital when the doctors delivered the horrific news to the parents. We all hoped for a miracle that day and prayed we would get word that the child had been revived, but deep down I knew it was too late when we got there. As I stood at the scene processing everything that was happening, I knew I would break down later in the day. I'd seen enough death and sadness, so I knew it was going to catch up with me hours later. I called Duane to tell him what was going on. I was hoping to meet with him for a hug, knowing I would need to have a good cry like the times in the past. There's two things a cop needs when they deal with calls like this. One is to hug their own child as soon as possible, the other is to privately cry, preferably in another adult's arms.

Later that day, we were all called in to a debriefing. New policies were in effect in an attempt to help first responders cope with calls like this. We were all given a business card and told of some things we might want to be on the lookout for like trouble sleeping, moodiness, drinking more than usual and withdrawing from friends and loved ones. Honestly I thought the whole thing was a waste of time. This would have been my fourth death investigation of a child and I had accepted that dealing with these traumatic events was just part of our job. What was the point in all this I thought? I just wanted to get home to Sophia and hold her tight like I always did at the end of a bad day.

Trauma is a strange thing. Sometimes we can experience horrific things and move on, other times, minor traumatic events will seem heavier than others. Much of how it affects us depends on our personal experiences and where we are in our lives. Sometimes it's just another traumatic event that happens to be the one that breaks us. This drowning was my "one too many" or the straw that broke the camel's back as some would say. I started to dwell on all of the other deaths and faces of those I'd tried to save. I started to worry myself sick over what if it was my child? The pain became too much to bear.

Strangely enough, I couldn't recall every adult death investigation I'd ever done or any details of them but I can recall every single detail of the four children I've responded to that died.

The one that always troubled me the most, was a 13 year old girl who hung herself in her family's garage. When we arrived, we were met with a mother, father and older brother, all of whom were hysterical and crying. The family was physically fighting with each other and screaming in a foreign language as they were trying to cut her down. Firefighters arrived almost at the same time we did and as they tried to assess the girl, we pulled the family into the house so they didn't have to witness what was going to happen next. I don't care how tough someone is, it doesn't get any worse than seeing a child take their own life.

While in the house, the mother asked to use her bathroom because she felt ill. More than understandable in a situation like this, so another officer agreed to let her go. The next thing we knew, the mother walked out of her bedroom holding a gun to her head. There were about six of us there, we all pulled our guns out of instinct. We were begging her to put it down but she just kept screaming and crying in a language we didn't understand, nor did we need to. Regardless of where someone is from or what language they speak, the grief from the loss of a child is universal and any mother from any corner of the world would grieve the same. Not a single one of us wanted to shoot her, we just wanted her to put the gun down so the rest of her family wouldn't witness it or suffer any more than they already were. Two officers were holding back the brother and father as they tried to get to the mother. The last thing we needed was a fight over the gun and for someone to accidentally get shot. Finally one of the officers managed to develop a dialogue with her and she eventually put the gun down. That was a call I will never forget. I made up my mind that day that I would never let a distraught family member out of my sight until I knew it was safe to do so. They could get as angry as they wanted with me and that would be okay, at least I would know I was doing the right thing.

As I drove home after the crisis debriefing, I was reminded of the little boy from the Netherlands that drowned early in my career. I thought of Sophia and how devastated I would be if anything ever happened to her. I called Duane to see if he could meet me that night because I just needed a shoulder to cry on and I didn't want Sophia to see me like that. She had already been through enough that week with losing Max and Grandma being in the hospital.

Duane told me I should call someone else because he wasn't good at that stuff. He was short and in one of his asshole moods. I got frustrated and expressed my disappointment after all I had done for him. I was pissed he couldn't find the time to at least give me a hug when I needed it most. I was also dumbfounded, typically he would leap at the chance to see me and I had just lent him $1700.00 to get his truck fixed.

After I got home, I learned a dog we were pet sitting peed on my bed and ruined my memory foam mattress. *What else could go wrong today* I thought to myself as I just began to cry. I headed to the store to buy a new topper because I couldn't sleep without it. As I drove by the mechanic shop Duane and I had used for his truck, I saw Duane standing there talking to the mechanic. I stopped to ask him what was going on and he pushed me away as if I had the plague. I was completely and utterly confused. The money I had just loaned him was to be repaid by him doing some work on the house. *Why was he being such an ass?* I thought.

After being humiliated, I got in my truck and continued on to the store. On my way home as I drove back by the shop, Duane was gone. I stopped to ask the mechanic what was going on and he said Duane had come to pick up the truck even though it wasn't ready yet. I asked him how Duane got there and that's when he told me "some woman he's been hanging out with. You didn't see her in that red car?" That's when I knew, he had found another woman to take care of him and all I was to him was a bank account. My sorrow turned to instant anger and all hell was about to break loose.

I went to his house and there sat his truck. Duane and the girl were nowhere to be found. I called his mother to see what she would say and she told me the truth as she apologized for her son's behavior. She thanked me for all I had done and put up with from him and said she understood that I had to look out for myself.

I called a friend and she took me to get Duane's truck. I drove it to a friend's property where it would be parked until he paid me the money he owed me and refinanced the truck without my name. It took a while, but eventually I finally broke even. It took me nine months to get his property out of my house because legally I couldn't throw it out. An attorney friend wrote a letter directing him to move his belongings out or he would be charged $500 a week for storage while an eviction process began. I really wanted to avoid filing for an eviction because it was expensive and completely ridiculous at that point.

In the weeks following the baby's death, I spiraled out of control. I could barely have a conversation without breaking into tears, I had withdrawn from all of my friends, I would just sit with my mom who was in a rehabilitation center at this point and had no recollection of everything that had happened over the past several weeks. Sophia was carrying the burden of trying to cheer me up and get her mom back and no matter how hard I tried and how much I wanted to feel better, I just couldn't do it. I felt like some sort of demon had reached from the grave and was pulling me down into the darkness. I was tired and I didn't have the strength to pull myself out. I didn't understand how with everything I had been through in my life, this stupid break up, that I wanted, that I knew was best for me, was bringing me down so hard.

Little did I know, it was much more than the breakup bringing me down. In a matter of three weeks I was seriously contemplating suicide for the first time in my life. I was in physical pain from the grief I was experiencing, I felt like a failure as a mother to my daughter, I just wanted to quit because I didn't have the strength to pick myself up anymore. I knew better than to tell anyone what I was thinking, they would for sure take my gun and put me in a mental hospital.

I was working for a Sergeant at the time I couldn't trust as far as I could throw him. It was as though he was looking for a reason to jam me up at work. I hated him. That's the thing about this job, we sign up knowing we will see horrific things and witness the worst in humanity. We know it's dangerous, we know some people will hate us just because of the badge we wear, but our desire to make a difference in this world and to protect the innocent keeps us going. What we do not expect however, is for our leaders to fail us when we need them most. Ever since the media began to bastardize us and our leaders began to care more about how we are perceived instead of how we really are, suicide rates in law enforcement have skyrocketed. We don't trust anyone, not even those we should because of how we see others treated by their leadership. Our leaders have become politicians and care more about themselves than having the guts to do the right thing even if it's not popular.

As I struggled with my pain and how I wanted to end it, I worried what would happen to Sophia if I did "check out." I knew the pain it would cause her, I had witnessed it too many times over the years, but the pain was killing me and I didn't know what to do.

It was a Sunday morning and I was on my floor sobbing uncontrollably in the fetal position. Sophia was visiting family in another state and I was alone for the week. Not a good thing for someone in my state of mind, but I sent her away because I didn't want her to see me the way I was. I thought it would do her some good to get a break from all the depression. As I laid crying on my bathroom floor, I remembered the debriefing we had. I found the phone number of the team leader and I called her. I knew she had been through her own fair share of trauma, so I took a chance in trusting her, I was desperate. I explained what I was feeling and that I couldn't get myself together. I did not mention the thoughts of suicide and I think she knew better than to ask. I would have lied anyway. She told me to go to an urgent care and explain to a doctor what I had been through. She said they could give me something to calm me down and help stop the crying. So I did.

I went into the urgent care and caught them just before they closed. They were very kind and understanding of what I was going through. The doctor wrote me a script for Xanax and suggested I follow up with a counselor.

The next appointment I made was with my OBGYN. I was convinced my problems had to be hormonal. I insisted he run labs on everything for me, which he did, but at the end of the appointment, he made it clear to me, as I sat in his office sobbing, that I was suffering from clinical depression. He wrote me a script for an antidepressant. Three days later I called his office and told his nurse to tell him to take those pills and shove them up his ass. I was hysterically crying and screaming at her when I made that call. To say I was unstable would have been an understatement and to this day I will never know how I managed to hide it from work other than by avoiding others all together. All it took was for someone to ask me, "How are you doing?" and I would burst into tears before I could answer.

On July 4th, Sophia and I were supposed to be celebrating with friends. I showed my face and it was clear I wasn't in a celebratory kinda mood. I left Sophia there to have fun and I went home to crawl back into bed. Nothing was working and I was so deep into the darkness at this point, I could see no way out.

The next day Sophia was preparing to go away to summer camp and she was very excited about it. I was happy for her, but concerned about being left home alone for another week. I was on the verge of really going over the edge of doing something awful. I was starting to think about how I would do it and decided it would be best to look like an accident. I wanted Sophia to be taken care of and there was plenty of life insurance to make sure of that, so long as I went as the result of an accident.

Sophia left for summer camp and after returning home, I was frightened by the thoughts in my head. I needed help but couldn't trust anyone. I knew the consequences of expressing my thoughts with any of my peers. I called the only person I could trust, I called Zach. I knew he wasn't going to answer, he never did on weekends, but this time he did. The second he said "Hello", I fell apart. I was crying so hard I couldn't breathe. All he could do was tell me to calm down and talk to him. Finally I managed to form the words burning in my brain, "I just want to die" I said. To which he replied, "Whoa Whoa Whoa, don't go doing anything stupid. Talk to me." I just rambled on about how much I was hurting and how I couldn't take it anymore and how I was failing as a mother. He was quick to express he understood how I felt, but he insisted I was far from being a failure as a mother. He spent the next eight hours on the phone with me telling me I was the strongest woman he knew and that while what I was going through was tough, I was tougher. I certainly didn't feel that way, but he was pretty persistent. He promised me that he wouldn't tell anyone what I was thinking if I promised to get some help right away. He convinced me I could get counseling and my employer would never know what we discussed. I eventually agreed.

The next day I called another friend whom I consider to be one of the smartest human beings I've ever known. In fact, I've told him a million times he should write a book about his life. He's a true inspiration and has overcome odds and made something great of himself. Raul knows all about pharmaceuticals, so I called him for help. I hated the side effects from the antidepressant my OBGYN had given me and I needed to know what else I should consider. The last thing I needed was more vivid nightmares and loss of sleep. I was already exhausted from all of the crying I was doing. I trusted Raul completely and I knew he wouldn't steer me wrong. He made a suggestion for a medication a lot of veterans he knew had used and he felt like it would work for me. I called my doctor to ask for that one and he agreed to write the script. He also emphasized the importance of seeking counseling, which I planned to do, since I promised Zach.

I got in with a counselor and continued seeing him even though I didn't feel it was helping much. He seemed frustrated with me because I was a bit hard headed and not really open to his advice. I was frustrated because he didn't understand how his advice would affect my career.

Twenty

Angels Come in a Variety of Ways

Initially the medication did nothing for me, but it also wasn't giving me any horrific nightmares. Unfortunately though, it also wasn't allowing me to sleep, so I would take a Xanax at bedtime for the sole purpose of sleeping. Doctors said it wasn't intended as a sleep aid, but it certainly worked for me. Sadly, too many people abuse it and it's highly addictive, so unless it's taken responsibly, it's a huge risk for doctors to prescribe. After seeing the toll pain killers took on my mom in addition to other mind altering medications over the years, the last thing I wanted to do was go down that path, so I made sure to take them responsibly and follow the rules. Only bed time, not when driving and never with alcohol, not that I drank much anyway.

There was a huge conference going on at work and all of the SRO's, School Resource Officers, like me, were assigned to work security for it. The first few days of the conference, countless officers I hadn't seen in awhile would comment about how much weight I had lost. In addition to the 100 pounds I lost intentionally, I had stopped eating and had lost another 40 pounds. I was reaching the point of being too thin and people were noticing. Nonetheless, I just couldn't eat, I was nauseous all the time and the thought of food only made it worse. One old friend approached me, gave me a hug and simply said, "How have you been?" Just like that tears welled up in my eyes and I was struggling to keep it together. He knew right away, something was bothering me and he wished me well and walked away. That guy would later become my sergeant and the best one I would ever work for.

Wednesday came and I woke up feeling different. It was day 10 of the antidepressants and I almost felt human again. I drove to the conference and for the first day since the drowning, I was able to socially interact with people.

By the end of the day there were only a handful of us sitting at the table near the entrance. I was the only female. The guys were mostly talking about sports and the stuff they usually like to talk about, I was just sitting there watching the clock, waiting for it to strike 5 o'clock so I could make the hour drive home.

People think angels appear from a bright light with majestic wings spread wide open as choir angels can be heard singing in a heavenly voice, but that isn't always true, not in my experience. As we sat there waiting to go home, this guy walked through the doors pulling his luggage. He approached us and asked where he should register. One of the guys told him he was a little late, the conference started three days ago. That's when this incredibly good looking man, about my age, replied, "Oh, I'm here to present. I'm speaking tomorrow." He looked a bit familiar to me but I couldn't place him. One of our sergeants sitting there recognized him too. He was a retired SWAT Commander from another county on the other side of the state and had worked for years as a judge at the International SWAT competition. The second I heard that, I knew exactly who he was, because I would stare at him as he sat up in his tower judging. He was an incredibly handsome man.

As it turned out, he was also a very friendly, funny guy. There is nothing sexier than a good looking man with a down to earth personality and a sense of humor. He introduced himself as Tommy and proceeded to stand there with all of us shooting the shit and cutting jokes. He made a comment about the "old days" and I spoke up with my 2 cents. He looked at me and said, "Oh right, like you would remember the old days. How old were you when you started? Twelve?" I laughed and replied, "Well, aren't you my new favorite person." It was at that moment I noticed a corporal sitting to my left, kinda raising his eyebrows because he picked up on the flirting that had just taken place. Honestly I didn't, I thought we were just joking around. I have never taken friendly banter as flirting so I often miss the cues when it happens. Tommy and I made a few more flirtatious comments over the next several minutes and then the clock struck 5. Everyone at the table stood up and headed for the doors. I had parked on the upper level so I had to take an escalator up toward the main lobby of the convention center/hotel. Tommy had to go that way as well to check in to the hotel. As I stepped onto the escalator, he stepped on right behind me. Seconds later he said to me, "Are you going by a coffee shop in the morning?" To which I replied, "No, I'm sorry, I don't drink coffee anymore." He literally looked as though someone had just stolen Christmas from him. So being the smart-ass I am, I replied, "Unless you WANT me to go by one." I expected him to tell me "never mind" or something like that, but instead he said, "Would you? That would be great!" Well now look what I got myself into. I told him I had the later shift and wouldn't be there until 9 and he said that would be fine. *Great*! I thought, *Now I'm some guy's coffee bitch.* Truth is though, he was too cute to say no to, so I agreed. Then he commenced to give me the most complicated coffee order I'd ever heard. There was no way I was going to remember it. That's when he said, "Give me your number and I'll text it to you." That certainly seemed easier to me, so I gave it to him.

A few hours later I got a text, it just said, "What are you doing?" I was literally sitting in the emergency room with my mom who wasn't feeling well. It didn't seem like anything too serious, but more of a precaution. While waiting, I sat and texted back and forth with him for a couple hours. Then when it was time to go home, I told him I couldn't text anymore because I needed to drive. The next thing I knew, he was calling me. It was clear at this point I had been played and the coffee bit was just a way to get my number. Pretty smooth actually, not bad for an old guy. I didn't care though, it was nice to be flirted with by someone who was not only good looking but truly had their shit together. My ego desperately needed it.

The following morning I showed up with his coffee and texted him. He told me where he was lecturing and to come in. As I opened the doors to what I thought was going to be a lecture in recess, I realized it was the grand ballroom and there were at least 300-400 cops in there, all men. As I started to walk in and realized Tommy was in the middle of his lecture, I started to back out. Tommy stopped his lecture, pointed to me and said, "Hey everyone, I want you to meet the most beautiful officer I've ever met. She was nice enough to pick me up some coffee this morning." Every head in the room turned around and looked at me. There I stood in my uniform, holding his coffee and turning 1000 shades of red. Now understand, I'm a salty sailor and a veteran cop, not much embarrasses me, but HOLY SHIT! He sure did a great job that day. He made me walk all the way to the front of the room down the very long aisle through the audience and as he stepped down off the stage with his wireless headset microphone, he hugged me, kissed me on the cheek and said, "Love ya, Babe." Everyone in there heard him. They also heard my reply, in the most sarcastic tone and with the dirtiest look possible, "Love. You. Too."

As I turned around to leave, it was painfully obvious how badly he had embarrassed me and every guy in there knew it. They all got a good laugh at my expense and chuckled as I walked by.

As I exited through the ballroom doors, I just stood there asking myself what the hell just happened. One of the sergeants looked at me and said, "You OK?" to which I just nodded my head because for once in my life, I was speechless.

A few minutes later I got my wits about me and texted Tommy, "I'm going to kill you for that." As soon as his group went on a break, he found me. Thanked me for the coffee and asked me to have lunch with him. The sergeant who knew him from the SWAT competition was quick to speak up and say, "I'll make sure she is available." When Tommy walked away, the sergeant commenced to tell me this guy was highly respected and I'd be a fool not to get to know him. It didn't take much persuasion, I did think he was pretty spectacular.

Hours later, Tommy came, the sergeant made sure he had my post covered and off we went. Tommy was in a hurry because he had run over on his lecture and he had to rush to be ready for his "break out" session starting in 45 minutes. We rushed to the cafe, grabbed some sandwiches and headed for the smaller lecture room where he would present next. We really didn't get to talk much as he scrambled to get his laptop working and presentation ready. He asked me if I could stay for a little while and listen to his lecture. The sergeant told me not to worry about hurrying back, so I told Tommy I could stay for a few minutes. I was kind of curious to hear what he was lecturing about.

As conference attendees began to stroll in, most of them had sat through his morning session. It was funny how they assumed Tommy and I were a couple based on how he addressed me in front of the crowd and he wasn't helping much as he introduced me as his wife to everyone. I'd never met anyone like him before. He was hysterical, yet convincing when he spoke.

Once everyone was in their seats, I took a seat in the back as I was still nibbling at my lunch. I also wanted to be able to get up and leave quietly without causing a disruption. Tommy was speaking about how many cops die by suicide every year. This topic I was not mentally prepared for. If he only knew how close I was just a couple weeks prior. I was instantly uncomfortable with the topic. This was hitting way too close to home and I didn't want to show any emotions.

Tommy started his lecture by telling a very personal story. He said he hadn't realized how years on the job had taken their toll on him. He talked about all the SWAT calls, dead bodies and political stress that came with the job. He talked about how many friends he had lost on the job and how he couldn't stop asking himself if he could have done anything better that would have changed the course of fate.

He got extremely serious and showed a side of himself I hadn't seen in the last 24 hours. It was raw and emotional. He spoke of how the job ended up costing him his 30 year marriage because he was never around and always gone. When he lost the support of his wife and began to realize his daughters would soon be leaving the nest, he felt alone. He found himself crying all the time and he described the same darkness I had been experiencing. He said all of the traumas he'd ever experienced kept playing over and over in his head and he couldn't make them stop. I swear, for a minute, it was as though he was inside my head because this was exactly what I had been going through all summer.

Tommy credited friends who got him help, he refused to go to the place where we, as cops, take everyone else. We'd rather die first before going there. Tommy walked into his captains office, took off his gun and badge and placed them on the desk. He told his Captain he needed some time away before he did something bad. He described how difficult it was as a SWAT Commander to make that move, it was his whole identity, but friends convinced him he couldn't help others if he didn't help himself. After getting a thorough medical and psychological work up, it turned out his depression was part PTSD and part hormonal. Once everything was addressed and treated, he began to feel better and returned to work. Ultimately he retired two years later and now spends his time telling his story so that he can help others before they use a permanent solution to what is often a temporary problem.

It took every ounce of strength I had to choke back tears as Tommy spoke. I had so much respect for him to tell that story because cops don't do that. We don't show weakness...EVER...for fear we will be sent off for a psych eval and then lose our jobs. The fear is real.

I waited until it was an appropriate time to get up and leave, I walked straight to the restroom and just trembled as I cried. Who was this guy? If I didn't know any better, I would think God stepped in and made this all happen just to send me a message. It was too ironic to be a coincidence.

I returned to my post and when Tommy went on break he came to see me. He asked me if I had plans for dinner and I told him, "yeah, I have to go home." He asked me to stay and join him for dinner and I laughed as I reminded him I was wearing a uniform and dinner and drinks are kinda frowned upon in my attire. He insisted I join him for dinner and asked if I could go home and change to which I told him I lived an hour away.

The sergeant overheard this conversation and stepped in to say, "Donna, why don't you leave a little early and go buy something close by?" I looked at him as though he was crazy. I told him the tourist district was too expensive and he asked me where I liked to shop. I told him my favorite store and he was quick to look it up on his phone and give me directions to a suburban area nearby. So I guess I had a date. Tommy told me when he would be done and we worked out a time for me to go buy some clothes and come back. A female deputy from another county heard about this and offered me her room to change and get ready. It was almost a team effort to get us together and it was extremely comical.

I found a cute outfit and even had time to get my makeup done at a cosmetic store. When I told the staff what I was doing, girls stepped up to do my hair and everything. I kinda felt like Julia Roberts in *Pretty Woman* only I wasn't a hooker, I was a cop.

When I got back to the hotel, I changed clothes and met Tommy. There was a convention meet and greet happening in the ballroom and we had to stop by there first before going out. There I met police leaders from all over the state of Florida as well as officials from the state capital. Tommy introduced me as his wife to everyone and it was hilarious to see the look on their faces as though they were surprised. Clearly he had an established reputation as a bachelor. Initially I kept correcting him, telling people I wasn't his wife, then after a while, I gave up and just went along with it. He seemed so entertained by leaving them perplexed that I just let him have his fun. I genuinely didn't care, it was too ridiculous.

After the party, we walked to the entertainment district which was just down the street. We had an amazing dinner and drinks and we never seemed to struggle to find things to talk about. I told him how much I respected him for speaking so honestly about what he had experienced and he opened up more about how difficult it was to start over after being married for so long. I never worked up the courage to tell him what I had been through, I just wasn't ready yet. I hoped that eventually I would be, but the fear of being seen as so broken and vulnerable was too strong, so I didn't say anything, I just listened.

By the time we were done having dinner and drinks, it was late and I couldn't drive home. Tommy had a suite at the hotel that had two rooms. He invited me to spend the night there since I had to be back so early in the morning. I found a babysitter for Sophia for the night, so I was able to accept his invitation. I've never been one to kiss and tell, but I will say I genuinely enjoyed being treated like a queen for the night and he remained a perfect gentleman. It was exactly what I needed.

The next morning when I went downstairs, my coworkers figured out I had stayed there for the night. Rumors flew and I honestly didn't care. The women were driving me crazy for details and I wouldn't share other than I had a great time and he was a perfect gentleman. I don't care what anyone says, women are far worse when it comes to "locker room" talk than men are. When the day ended and Tommy was leaving, he made sure to stop by and say goodbye. We stayed in touch regularly and have gotten together a few times when he is travelling on my side of the state. We have become the greatest of friends and while he now knows about all of my demons, he is always there for me when I need him. I knew from the beginning a relationship was never going to happen, not when we lived three hours away from each other. Nonetheless, to say he holds a special place in my heart would be an understatement. He alone got me to see I could do so much better than what I had been settling for and he didn't even know it.

I continued with counseling which helped reaffirm what I had just discovered and my counselor seemed genuinely happy for me, although he was concerned why I was hesitant to enter into another relationship with anyone. I felt like I had done so well on my own, getting where I was, I didn't want to mess that up again. I was still waiting for Duane to get his crap out of my house and I was done being nice.

As I attempted to juggle all of the drama Duane had brought into my life, the stress was still getting the best of me. It was August,

2017 and I was in active assailant training as we prepared for the new school year to begin. I had gone through this type of training countless times over the years and while the protective face and head gear would bother me tremendously as we moved through scenarios, I had always been able to work through my claustrophobia. My best guess is due to all the stress I had been under that summer since losing the toddler in the drowning in addition to everything else, that day was too much. I worked through six out of the seven scenarios successfully, but the seventh scenario went longer than usual and I found myself the only survivor in my group as I engaged in a gunfight with the shooter. Fortunately I was able to fight through my fears to complete the scenario, but as soon as the instructor yelled "End Scenario!" an enormous wave of fear overcame me and all of my adrenaline dumped as the panic attack came at me like a freight train. It happened so fast I couldn't work through it. I was struggling to get the headgear off and before I knew it, I couldn't breathe, the tunnel vision kicked in and the hyperventilating began. All I could think of was ripping all my gear off to include my bullet proof vest and shirt. The anxiety overcame me and a side of me no one had ever seen came out. That became very obvious as I began cursing "Fuck this Shit!" as I threw my helmet across the room and began to run out of the building ripping off my gun belt, vest and anything else I felt was smothering me. I was overcome with emotion and fear as I cried and desperately fought to get control of my breathing.

Both of my lieutenants, a male and female, chased me outside and found me hiding in a corner away from everyone and everything. I just wanted to be alone and hide. The female lieutenant did a much better job of calming me as she told me I was OK and that I finished the scenario by winning. The male lieutenant was a bit more shaken and wanted to call me an ambulance. That was the last thing I wanted, more attention. I just wanted to disappear and breathe. Eventually with her help, I did, however I was incredibly embarrassed and humiliated over the whole ordeal.

When word got to my sergeant what happened, he was directed to check on me to see if I was OK and find out what happened. Not once did that asshole ever ask me if I was OK. Instead, he held the incident over my head for four months and used it as a reason to deny me opportunities. I attempted several times to talk to him about the matter and explain to him that there had been a letter in my human resources file about being claustrophobic for years, but he told me he wasn't ready and to stop bringing it up. That day never came.

It was around this time I got an unexpected message from an old friend from Panama. Reagan and I had been friends on social media for a while and we shared a passion for German Shepherds. It seemed his female had a litter of pups and he kept one for himself to add to his pack. He saw I was struggling, judging by things I would post online and he knew getting a dog would help heal me, even if I didn't know it myself. He spoke with his wife who agreed four dogs in one house with kids was too much. It seemed Larry the pup needed to find a good home. Reagan told me that he wanted me to have the dog because he knew how well he'd be cared for. He also knew that I needed the dog as much as he needed me. At first I said no. I wasn't ready, I was still grieving the loss of Max. Reagan told me to give it some thought, I was the only person he was willing to give Larry to.

I hung up and discussed it with a friend who thought it was a great idea. Then I discussed it with my counselor who strongly recommended it. After several days, I called Reagan and asked him if he could wait until Labor Day for me to come because I wanted to drive to New York to pick him up. Reagan had offered to put Larry on a plane, but I couldn't do that to him. The only way I was willing to let him travel was by car, with us.

Over the next couple weeks I prepared the house for a puppy German shepherd. I had already fallen in love with him based on the photos and videos Reagan had sent me. Sophia and I both agreed we didn't like the name Larry, so we came up with a name we both liked. We named him Saint Michael after the Archangel who was the Patron Saint of law enforcement and military. It seemed perfect since Reagan was a retired SEAL and I was a cop. We would call him Saint for short.

It was a Friday afternoon and there was a hurricane brewing out in the Atlantic. My neighbor and close friend Claire said she would make the road trip with me and we could stop in Philly to stay with her mom. Reagan's house was only a few hours further north of her. We left on a Friday right after work and drove straight through to Philadelphia. We arrived at 10 am the next morning, exhausted. All of us crashed to get some sleep. Later that night she took me out to enjoy a real Philly Cheesesteak and we met some of her childhood best friends. Sunday morning, Sophia and I got in the truck and headed for Poughkeepsie, NY. It was a quaint drive, but I could tell it was much different than living in Florida. There were liberal parades and protesters along the way in little towns we passed through. I guess they were still pissed off about losing the election and couldn't accept that Donald Trump was still their President.

We finally arrived at Reagan's house. The weather was wet and chilly, clearly fall was well on its way in New England while we were still dying of heat in Florida.

Saint was less than excited about meeting us, Reagan said he knew it was his turn to leave the nest. I could tell Reagan's kids weren't happy about seeing him go and he told me they were mad at him for giving him away. It was clear that Reagan was doing this for Saint's sake and he had already attached to him, but he had to go and Reagan was happy he was going to a good home.

After visiting with Reagan and his family for a couple hours, Sophia and I left. Reagan had to physically carry Saint to the truck to put him in and it wasn't easy. At barely 7 months old, he was already 65 pounds and a huge ball of fur. Unlike any German shepherds I was used to seeing in Florida, Saint was a long-haired shepherd and he was absolutely beautiful.

Sophia sat with him and comforted him the best she could in the back seat. He clearly wasn't accustomed to car rides and did a great job of showing us as he vomited all over the back seat and floor.

Once we got back to Claire's Mom's place in Philly, we cleaned out the truck and prayed Saint would do better the next morning as we drove back to Florida. The poor baby was scared to death and I could tell that if I let him off leash even for a second he would take off trying to find his way home.

Saint wouldn't eat, drink or use the restroom the entire time we had him. It wasn't until we made it halfway to Florida the next day, he finally peed. I just kept hoping he would eventually learn we just wanted to love him and give him a good home.

Meanwhile, as we made the long drive home, Sophia continued to bond with Saint in the back seat as Claire and I monitored the hurricane headed for Florida. We knew by the time we got home and got to the stores, there would be nothing left on the shelves. So we picked up supplies as we drove south. That turned out to be a very smart idea.

When we finally got home in the early morning hours of the next day, we got some sleep and then started preparing for the storm. I helped Claire and she helped me. We got sandbags and boards put up. Poor Saint was so confused. Where was this place we had brought him to where it was hot, humid and people covered their windows?

The storm came and we sustained minor damage. The worst was losing my fence again and not having power. After four days of no power, we couldn't take the heat anymore so we loaded up and drove to a friend's house where there was power. No sooner did we arrive at her house two hours away, our neighbor called and said our power was back on.

The next morning we returned home to start the clean up. The refrigerator was cleaned out and all of the spoiled food was thrown out. I got to work finding a fence company to replace the fence that was blown away. Saint was so confused and scared but so sweet...until he saw another animal. He wanted to kill anything that had more than two legs. It was time to get to work training our new pup and the adventure was just beginning.

Twenty-One

Man's Best Friend

First things first, finding a dog trainer to help train a very big seven month old German shepherd puppy. Fortunately Saint was already house broken, so at least we didn't have to contend with that. Other than that, he had no concept of any other commands. Every time we tried to go out the front door, he would run. He thought it was a game of "Catch me if you Can." I was so worried he was going to attack someone or their dog while they were walking them or even worse, I was terrified he was going to get hit by a car. He had no concept of "Saint NO!" or "Come!". "Sit and Stay" was a joke. It was as though he had a motor that he just couldn't control and there was no wearing him out. His energy was boundless. Whenever we would walk him, he would go nuts if he saw another animal, dog, cat, squirrel, it didn't matter, if it had fur of any kind, he wanted to kill it and he was so strong, neither my daughter nor myself could hold him back.

I met with a local K9 officer I knew and he offered to help me with Saint. Brian was an amazing canine handler and folks said he was the master. The night Saint and I met up with Brian, it was as though he was a different dog. Not once did he act up or show his ass. He didn't go after Brian's dog, he minded and was a total Saint. He made me look like a complete liar. Brian laughed and told me what an amazing dog I had and he didn't see the problem. I was so confused. Why was he perfectly behaved around Brian but not for me? This is an answer I would eventually come to understand. Often training a dog isn't really training the dog. It's the human that needs to be trained, and I desperately needed training.

Brian and I worked a little together but our work schedules just seemed to clash. So I enrolled in a night class at a local dog boarding facility. There we stood out like a sore thumb. Classes were held on Wednesday nights and when we showed up for class, it felt as though we were in "munchkin land." Every dog there was a little ankle biter and Saint got the memo that we were going to an hors d'oeuvres party. Through the entire class of EIGHT WEEKS, Saint showed his ass and spent every minute of the hour we were there trying to bully all the little dogs. It was so embarrassing and it was as though I had brought Dennis the Menace to school.

Finally at the end the instructor suggested I take Saint to an expert. He recommended a guy who claimed to be the "behind the scenes" guy for a famous Hollywood dog trainer. I was desperate and I was willing to try anything.

I drove over an hour to get to this guy's house, paid $250 an hour for him to tell me, "your dog is too aggressive and if he isn't in the hands of an expert, he's going to hurt someone." So basically he was telling me to give my dog away to an "expert." I cried all the way home. Saint meant so much to me and I could tell he had a big heart. We had developed a strong bond, but I needed him to behave so we could do things together without him terrorizing everyone and everything in sight.

I gave up for a while and just kept Saint in the house or we would go places to run and play where there weren't any other people or animals. That worked for a while, but wasn't exactly what I hoped for. I needed to be able to take him out in public to dog friendly venues. I lived in a town where restaurants have outdoor seating and are very dog friendly.

I didn't realize it at the time, but Saint was quickly becoming my rock and I found I needed him. He slept with me and always had to be touching me in some way, even if it was just his paw. I was starting to sleep a little better and when I would have a nightmare, he was there to calm me when I awoke. Little did I know at the time, this dog was saving my life and distracting me from everything in my head that had broken me down.

One night while meeting with Brian and one of his coworkers, they suggested I look into a program at the local sheriff's office where they train dogs to be therapy dogs. They thought Saint would make a great therapy dog because of his friendly nature with humans. I explained to them I wasn't so sure about him around other dogs though because of his aggression and they assured me, with the program, he would get over that.

I was put in contact with the guy in charge of the program and told him all about my boy. As luck would have it, there was a last minute slot in the upcoming fall class due to a cancellation. All I had to do was get my agency to sign off so I could go. I just knew they would since it was completely free. Our division leader was a big supporter of training and he had told the unit he would never say no to training, especially if it was free of charge, so I saw no reason why they would say no.

I quickly put together a training request package and explained the benefits of having a "therapy dog." I included scientific research, other law enforcement agency policies, news articles from across the country about what an asset a therapy dog could be to not only the agency, but the community as well. I was on to something great and I had found something that would give me purpose.

Therapy dogs were being used all over the country as a means of comforting victims of violent crimes, especially children. Scientific research was showing that the human brain releases endorphins whenever a person interacts with a dog. These endorphins bring peace and happiness, which is extraordinarily helpful when trying to interview someone who's been traumatized from a crime, especially sex crimes.

Therapy dogs were already being used in our local courtrooms to keep victims calm while they testified on a witness stand about the details of their traumas. They found prosecution rates were much higher and more successful when a therapy dog was used during initial interviews as well as in courtrooms. There is just something about a dog that allows a person to stay calm and speak of horrific things they normally would be too scared to speak of.

All of my research showed me why so many veterans were using therapy dogs as service dogs. Dogs were being credited with saving lives because they gave their owners a reason to live. They sensed when their owner was having a bad day or when their anxiety was on the rise and could distract them from their thoughts.

Saint had been doing this for me all along and I just didn't know it. I was having fewer anger outbursts and I was beginning to feel like I had purpose. Even though my brain already knew I did, I struggled with FEELING as though I did because I lacked the positive chemicals in my brain.

Unfortunately, I had a sergeant at the time who didn't appreciate my opinion. Quite honestly, he was just a dick. He had it in for me because he had been chasing my best friend at work and didn't stand a chance with her. Instead of moving on, he took his butt hurt ego out on me. He was an incredibly insecure man but walked around as though he was some war hero. He was a retired Master Chief in the Navy and acted as though he had some impressive background. He managed to fool a lot of people, namely our captain, but after doing some digging, I learned he had spent most of his career as a reservist and while he may have deployed a few times, he would never measure up to those who served their time on active duty for 20 years. I absolutely hated him. He belittled me every chance he got. He knew I had been through hell and not once did he ever offer a kind word or show concern for my well being. Instead he held it all against me and literally called me "thin skinned" and "weak."

I turned my request into him which had to be approved by the captain. He would hold on to it for a while and then give it back to me saying the lieutenant had questions before it was passed on. It was only a month before the course would start and my package was sitting on desks collecting dust.

After answering all the questions in a timely manner as they came my way, my request sat, unanswered, on the lieutenant's desk before I was denied on the Thursday before the class started. I was furious! There was no logical explanation for denying my request. I provided them with all of the data showing what an asset a therapy dog would be to have not only for our officers, but for our community.

I requested a meeting with the captain. Unbeknownst to me, the captain had been completely snowed by my sergeant who had poisoned the entire idea. Nonetheless, I got my meeting and when I asked "Why my request was denied," the captain said, "Didn't your sergeant explain that to you?" To which I replied, "No Sir, he didn't. He just said you denied it and I want to know why?" Then I reminded him how he told our entire unit he would never deny free training. To which he replied, "well this program won't benefit the agency, because we don't have a program like this in place." That is when I explained to him how many other agencies in our area did and how we were behind the times not having one. I further explained my hope was to get certified and start a program for our agency. I had already done most of the work, I just needed to take the next step. The captain denied my request but told me to keep working on it because it sounded like something that might be a good idea for the future. Without the class, though, I was pretty much dead in the water. I was furious. So I looked right at my sergeant and went off. I complained how he never communicated and how I didn't trust him to ever do the right thing when it involved me. I reminded him of an incident he was supposed to look into months before and how every time I tried to ask him about the status of it, he would yell at me and tell me not to bring it up again. The captain looked very surprised and said he needed to excuse himself, but that he would send in the lieutenant to handle matters.

When the lieutenant entered, I threw my sergeant right under the bus. He had been told months prior to sit down with me to discuss an incident where I had a claustrophobic panic attack during a training exercise. My claustrophobia was well documented in my human resources file from early on in my career. I was assured by everyone I spoke to from captains to training supervisors that I had nothing to worry about since it was documented. Anytime something, like a full face helmet, is placed on my head, I panic because I felt as though I couldn't breathe. I blamed it on the scuba training incident that happened when I was in the Azores, which certainly didn't help my claustrophobia, but I knew the real reason was the rape, I just didn't want to discuss that, so I kept it to myself.

The sergeant was furious with me for diming him out for not doing his job. The bosses didn't want me in trouble for what happened, they merely wanted him to talk to me and make sure I was OK, but he never did. He led me to believe I was under formal investigation and might be in trouble which was one more thing for me to worry about.

Later I would learn he was so pissed at me for standing up to him in front of the bosses, he took it upon himself to say I was "unstable" and was trying to get my gun and badge taken. Thankfully I had friends in high places who knew the truth and could see what he was up to. They intercepted and managed to get him shot down. Nonetheless, I was forced to go see a psychiatrist about my claustrophobia. He understood that the panic attack happened four months prior, but what he couldn't understand was why, if my supervisor was so concerned, it took so long to refer me for an evaluation when I had been working full duty since the incident happened. I told the Psych about the scuba accident and explained that was why I was claustrophobic. We discussed my career which was on it's 18th year with no discipline or any marks. It seemed pretty obvious he understood our visit was a waste of time. I made sure to mention my sergeant had scheduled that appointment during what was our squad Christmas party so I couldn't go. Not a coincidence I was sure.

Needless to say, after throwing my sergeant under the bus to the captain, because quite honestly, I couldn't tolerate his bullshit anymore, my request to attend the therapy dog training was denied. I was, however, told I could attend the training on my own vacation time. I thought about it for a moment and I felt so strongly about the benefits, that I agreed to do just that.

On the first day of school, I walked in with my eight month old German Shepherd puppy. He immediately showed his ass when he saw the other dogs there. A female instructor was not impressed as he was too high strung and too immature. Nonetheless, the guy in charge told her to give us a chance. We sat in the back of the room and Saint eventually calmed down. He didn't like it when another dog would look at him and he felt he was being challenged. I did my best to keep him under control but it wasn't going well.

The instructor was showing videos of how therapy dogs could be beneficial during interviews with sexual assault victims. The point of the video was to show the difference in the victim's demeanor both with and without a therapy dog so we as students could see the benefits.

I didn't even realize it, but every time we watched a video and a victim started to describe their assault, Saint would begin to act up. He would stand up, demand my attention, jump in my lap and whine. It was incredibly distracting. As a result, my stress level spiked and I was embarrassed as I tried to gain control of him. It was as though our anxiety was just feeding off of each other and we were getting nowhere.

Finally an instructor pulled me out and asked me to put Saint in one of the kennels because of the distraction. I was sent back into the classroom where I fought back tears. I was shaking and it was taking every ounce of strength I had not to completely lose it in there.

Finally lunchtime came and everyone was dismissed. The lead instructor, who was a combat veteran and seasoned sex crimes detective asked me to stay. He took Saint and I into a private area where we could talk. James asked me a question I never saw coming. He could see my anxiety level was incredibly high and that I was struggling to keep it together. He was smart enough to know there had to be more to my anxiety than just my dog misbehaving, he had read me like a book.

James said to me, "Didn't you tell me you served in the military?" To which I shook my head yes as I fought back tears. James knew me for all of about four hours and he asked me the one question no one in 28 years had asked me. He said, "Did something happen to you when you served?" I looked at him somewhat stunned, but unable to hold back my tears any longer and I just began to sob sitting on that couch. I couldn't breathe, it felt as though someone just pulled the plug from the dam I had built and now all the water was pouring out and I just couldn't hold it back anymore.

As it turned out, James had severe PTSD from his time in the military. His therapy dog started out as his personal service dog who was training to wake him from nightmares. James was incredibly bright. He used his GI Bill to get his PhD in Sociology and after teaching his family dog to wake him from nightmares, which subsequently saved his life due to the heart damage the nightmares had created, he decided to specialize in this area. James was the one who created the therapy dog program which was a one of a kind course and the only place in the country that taught officers how to utilize dogs as tools to help not only other officers who experienced traumatic events, but their communities as well.

James went on to explain to me that while Saint didn't have the right temperament for a therapy dog, at least not yet due to his immaturity, he would however make a phenomenal service dog for me based on how strong our bond was and how in tune to me he was. He asked me a series of questions about my fears and behaviors. He said Saint could be trained to help me with all of the areas I struggled with and because he was so bonded to me, he would be really good at it. What James picked up on was that every time the videos would show a victim describing their trauma, Saint would act up trying to distract me from my anxiety which was increasing because of what I was watching. Saint was trying to help me, not misbehave.

James asked me if I had any other dogs and I told him I did, but she wasn't very smart. He laughed and said this job wasn't about intelligence, it was about being calm and loving. Reluctantly I went home at lunch and brought back our other dog Snickers, a seven year old Lhasa Apso "Santa Claus" had brought my daughter when she was five. To my surprise, she was born to be a therapy dog and she sailed through the training the rest of the week.

Meanwhile, James continued to talk to me about everything I had been through and was continuing to struggle with. He asked if I had gotten any treatment from the VA (Veteran's Administration) and I told him that I had never reported the rape and nothing was documented. I explained to him that other than a close friend (Zach), I had never told anyone. James assured me I was not alone and that's when I learned the number. One in four women who served in the military during the 80's and 90's had been sexually assaulted during their time in service. He assured me I was not the only one who had served and not reported their abuse. Despite James' certainty, I was terrified of reporting the incident. I didn't want to be called a liar or humiliated. I had buried the matter so many years ago and I didn't want to revisit it or deal with it. James asked me some questions about my history of romantic relationships and I told him they weren't good. As a rule I pushed men away because I didn't trust them and the ones I had let in, just did more damage. I was done with men, I was broken and it was best I just leave myself on the damaged goods shelf. It was safer there.

James told me the next day they were going to watch more sex crimes interviews and given I had been a cop so long, he didn't feel it was necessary to force me to sit through that portion. He made a phone call for me to the local DAV (Disabled American Veteran) Office and insisted I go down there and file for benefits. I was extremely skeptical, but I did it because honestly, he didn't give me a choice.

After filling out my Notice of Intent to File at the DAV, I was directed to go to the local VA Clinic and tell them I needed help. I was extremely uncomfortable doing this, but James promised me it would be worth it. He told me to just go there and tell them I was a rape survivor and needed counseling. I couldn't understand how in the world I was going to get help for something that was never reported and not in my service records. Nonetheless, I did what I was told.

I arrived at the small outpatient clinic located near my home. I parked and walked in, barely holding my tears back. As I walked up to the window to register and get directions, the woman just asked for my VA Card and I told her I didn't have one. Then she asked for my social security number which wasn't in the system. She asked me if I had a job and income to which I said Yes. Then she asked me how much I made and if I had private insurance, so I told her. At that she said I wasn't eligible to see anyone there because of my current employment. I got frustrated and the tears began to fall. I was trying to explain to her what I was told to do and I couldn't get the words to form or come out of my mouth. As I stood there on the brink of a full blown melt down and panic attack, I saw it. A sign, pinned up on her cubicle wall right over her head. The sign had an acronym on it that said MST and what it meant, Military Sexual Trauma, followed by a hotline number. Without saying a word, I just pointed to the sign. She turned to look and when she turned back around and saw the look in my eyes, she very nicely said, "Have a seat, I'll be right back."

Within a few minutes I was directed to an area toward the back of the clinic. There I gave them my name and was told to have a seat, someone would be right with me. Sure enough, within minutes a nurse came to get me and escorted me to a doctor's office. My vitals were taken and I was asked some questions. After that, I was escorted to another area where a psychologist was waiting for me. There I described, as best I could, why I was there and who sent me.

That's when I found out about a program designed for women just like me. Many had kept their secrets for years and for whatever personal reason, were beginning to come out about what happened to them during their years in service. I was far from being alone. The statistics James had given me were accurate. One in four female veterans were victims of rape or sexual assault during their service and had remained silent for fear of speaking out. They explained to me I could be seen for as long as necessary under a hardship program designed just for us. In the meantime I was to apply for VA benefits which could take a long time.

After 28 years of silence, my secret was out and there was help for me. I was terrified of going through the process, the constant fear of someone not believing me was real. The fear of being called a liar or being seen as weak was often paralyzing. I had put on the facade of being strong for so long, but my walls were crumbling down and I just didn't have the strength to keep them up any longer.

Around the time Saint was turning one, I felt as though I was losing control. His separation anxiety from me was through the roof and he was literally eating my house. There were holes in the drywall below the front window, on the corners of the hallways and he was incorrigible if I tried to take him anywhere. I was a wreck because I felt I was losing all control. Then I heard about a K9 trainer that trained police dogs for my agency. I heard Matt was the best around, but I didn't know he trained personal dogs. I called him, explained my situation and he invited me out that Sunday to join his class.

That Sunday we showed up and Saint immediately tried to show the other dogs who the Alpha dog was, only THIS time, the other dogs weren't ankle biters. At least 95% of them were German Shepherds, just like him. The rest of the breeds were even bigger. There were roughly 30 dogs there divided into beginner and advanced groups. In a matter of a couple minutes, Matt put Saint in a position surrounded by other trained shepherds and he was quickly humbled and corrected. It was immediately understood that his attitude was not acceptable. I was taught how to correct him and while it worked for the time being, I couldn't help but wonder what he would do when it was just the two of us in public.

Saint was a quick learner that day, everything came easy once he figured out who the boss was. I couldn't wait to try what we had learned. That night we went to the village where all of the locals took their dogs to socialize. When Saint started to flex his muscle, I immediately corrected him like I was taught and it was a miracle. He looked up at me with his big brown eyes as if to say, "Sorry Mom." I was in awe! I'd found the best dog trainer on the planet. We continued to go every weekend and made lots of friends. Saint was no longer my pet, he had become my best friend and my protector. He slept with me and never left my side. If I had a bad day at work, he would know when I got home and he'd find a way to make me laugh. He was the best medicine anyone could ever ask for and became living proof of why a dog really is man's best friend.

Twenty-Two

Life Can Throw Curve-Balls

After promising Zach I would see a counselor, I had kept my word even though I didn't feel as though it was doing any good. The medications were helping, but talking to a stranger wasn't. He said he believed I was experiencing the symptoms of PTSD and I quickly shot that theory down. I didn't want that label, I was afraid they would take my gun and badge if they thought I was unstable. Granted, I have many friends who have PTSD from combat in the military and I knew better, but the stigma at the time was that if someone had PTSD, they were a loose canon.

When I started getting treatment at the VA, things changed, for the better. After explaining everything I had been through recently regarding Max dying, the abusive relationship with Duane, the baby drowning and my struggles at work, the VA took over my mental health care. They continued to keep me on the anti-depressants I was on and allowed me to continue with the Xanax for bedtime or unusually difficult days and triggers.

By this time, I was counting my blessings that I had woken up when I did. My relationship with Duane had been a hard lesson learned and I considered myself fortunate I was able to find the strength to get away and fight for myself. I've never judged an older man for being with a younger woman or vice versa since. I feel bad for those who are only loved for their money and I'm happy I'm able to support myself and my daughter on my own.

I continued to see a counselor after work on a weekly basis and talked to her about how I was feeling, my day to day struggles and how I managed them. My trust issues with men and people in general remained. I was put on a waiting list to go through a program called Cognitive Processing Therapy (CPT). It was a twelve week intensive therapy program designed for people with PTSD to help them revisit their trauma and work through it so they can have a healthier mental outcome.

Meanwhile, while visiting my counselor every week, I was told to give dating a try again. I was reminded that if I didn't try, I'd never succeed. I did want my daughter to see me in a healthy relationship so she wouldn't think all men were bad. Duane had really done a number on her and she was very angry with him for hurting me the way he did. I needed her to see that just because you get hurt, you don't quit. It was a lot easier said than done and I was really struggling with finding the courage to practice what I preached.

By this time she had figured out that Zach and I were much more than just friends. She could tell by the way we talked to each other that our relationship was more than that. She would hear us say "Love you" before we ended a conversation and to her, we needed to be together. I explained to her that wasn't possible because he was married and had a family of his own, but all she could see was how we felt about each other and she wanted me to be happy and she wanted him to be her dad.

Then hell froze over. I was at work one day on the gun range doing annual qualifications when I got a text message. Initially I didn't recognize it because it never got used. Nonetheless, I had programmed the only number I had for Damon, Sophia's father, as "Donor" in my phone. After all, that was the only title he was really entitled to at that point. She was just about to turn twelve when he reached out to me. The text simply said, "My wife and I will be in town in December and we were hoping we could meet Sophia and maybe spend some time with her." I literally stood there for what seemed like an eternity staring at my phone, trying to process who it was and what it said. It was surprisingly the same feeling I had the day I discovered I was pregnant.

I excused myself and made an emergency call to my counselor asking what I should do. Sophia wanted to know her father so badly, but I was scared to death he would meet her and then disappear again, only hurting her more than he already had. I knew his disappearance wasn't her fault, it was his, but she was only twelve and I knew she would take it personally if it didn't work out.

The counselor told me to make sure he had good intentions and not to allow her to be alone with him for a minute. I texted him back and asked him what his intentions were. Did he just want to meet her because he was in town and curious or did he want to build a relationship with her?

He told me he'd like to get to know her if I would allow it and I told him I would on three conditions. Don't make promises he couldn't or wouldn't keep. Be consistent and don't hurt her.

We talked on the phone as I drove home and I told him I thought it would be best to wait until the end of the week to call her because it was going to overwhelm her and I didn't want her distracted from school. He agreed and we set up a time for a conference call. It was the toughest secret I'd ever kept, even if it was only for two days.

Friday night came and I picked mom up on my way home. We were about to tell her the best news of her life and mom didn't want to miss it. When we got home, Sophia wondered why mom was at the house on a Friday night. That was not the norm for us for sure. She thought we were going to give her bad news like Grandma was sick and she started getting all worried. Mom and I assured it was nothing like that and to just relax until dinner was on the table.

Once we sat down and said Grace, I said to Sophia, "You know how you've been praying for a dad for so long?" She said, "yes." I replied, "Well, God answers prayers when he thinks it's the right time to answer them. That time has come and your prayers have been answered." Sophia replied with a statement that left me no choice but to laugh out loud. She said, "How could you be getting married? You're not even dating anyone." I laughed and said, "No, you're right, I'm not dating and I'm not getting married." The poor kid was so confused. That's when I said, "Your dad reached out to me and he's coming to Florida. He wants to meet you." It took her a minute to process the information and then to make sure she was right, she said, "Dad? As in Damon Dad?" I nodded my head yes and she instantly began to cry. She cried so hard that we all started crying. She was so happy. Then she asked when and for details. I asked her if she would like to ask him herself because if she was comfortable enough calling him, he was waiting by the phone. She immediately got my phone and called him via conference call so she could see what he looked like. They talked for a little while but it seemed awkward. Damon isn't the best at communicating, much less with kids. Then he told Sophia she had an older sister who was in college and couldn't wait to meet her. Phone numbers were immediately exchanged and soon thereafter, the girls started talking and getting to know each other.

We were all in awe how much the girls were alike considering they were half sisters and raised completely differently by two different parents. Nonetheless, they sounded alike, had the same likes and dislikes, same musical and artistic talents and the uncanny list went on and on. Sometimes I couldn't tell them apart from another room when they laughed.

After everything Sophia and I had been through that year, this was a welcomed event. Seeing her so happy made me happy, even if I was scared he would let her down. What if I was wrong and she found out I stood in the way of her knowing her father? I knew that was something she would never forgive me for and God knows I knew better.

Meanwhile I met a nice guy who worked in another unit at my agency. Neither of us had ever dated anyone at work, but we met via a dating site and after talking, realized we both worked for the same boss. We decided to keep things quiet and respected each other's professional reputations. He treated me well, but much like me, he had his own wall built around him. He had been divorced a very long time and was very comfortable living his own life. Like me though, he knew something was missing and he was hoping to fill the void.

After a couple months of dating every few weeks, we grew closer as friends. Neither of us were in a hurry to get into anything serious, but it was nice to go have dinner or catch a movie together. Then finally I got the call from the VA that my name was up on the waiting list for the CPT therapy.

When I showed up for my first appointment I met with the psychologist. He was a nice man and asked if I was comfortable speaking with him about what happened to me in the Navy. I told him I couldn't answer that as I was just meeting him and while I struggled trusting men, I had grown used to it because of my profession. We agreed to move forward and if at any point I didn't feel comfortable, he could refer me to a female if I wished. I agreed.

The first appointment wasn't bad. More of a "what to expect" appointment and I learned homework would be involved on my part. While I was less than thrilled about addressing my issues outside the VA on my own time, I knew I had to do it. I was tired of feeling the way I did.

Then over the course of the next few appointments, I was forced to revisit the rape and write down in detail what happened to me. It had to be handwritten for some psychological reasons. Something about the way the brain works and when a person hand writes something, their words have more meaning and they are more connected to their own thoughts.

On about the third or fourth appointment I made the mistake of calling Frank, the guy I'd been seeing, and pouring out all of the emotions I was feeling as a result of the giant gaping wound I had just ripped the bandage off of. Frank was not prepared for that. He was in this relationship more for the companionship and fun of it. He was not ready for me to dump a bunch of emotional baggage on him. That's when he said he didn't mind being friends, but he was hoping for a relationship that was a bit lighter and not so heavy. In other words, no drama and no baggage. Later I would learn he was carrying enough baggage of his own which was why he didn't want to take on mine. It wasn't meant to be and that was OK. We stayed friends with no hard feelings.

December rolled around and the big day Sophia was waiting for finally came. We drove to the city where her father and his wife were staying. They had spent the day at the beach and when we texted him to tell him we were there, they made their way through the hotel lobby. There he was, a t-shirt and beach towel wrapped around his waist. He had really aged since I saw him last. He seemed shorter than I remembered, older. He walked as though every part of his body was in pain and the anguish showed on his face, even if he was trying to hide it. Nonetheless, Sophia wrapped her arms around him as though he was the only man on the planet. Maggie and I, being the typical moms we were, both video recorded the entire thing. It was truly a very special moment.

After hanging out at the hotel with them, we all loaded up and drove back to the coast to our house. Damon and Maggie slept in our guestroom and while Damon and Sophia got to know each other, Maggie and I began a friendship. We went zip lining that weekend and spent some time at the beach. It truly was an awesome weekend. They invited us to join them for Christmas in Oklahoma and we accepted. There Sophia would get to meet her sister and her step brothers, Maggie's boys. Maggie was Damon's third wife and I believe she played a vital role in pushing Damon to reach out to Sophia. I was grateful to her for that because I knew how important it was for Sophia to know her father.

A few weeks later, Sophia and I spent Christmas at home with my mom and then after opening Christmas presents on Christmas morning, we headed to the airport to spend Christmas in Oklahoma.

On the morning of December 26th, we opened presents with everyone there. Maggie cooked an amazing Christmas dinner and the entire family, uncles, aunts, cousins and grandparents came to meet us. It was truly a magical Christmas and we all had an amazing time. Sophia and I repeatedly said to each other how nice it was, especially after the Christmas we endured the year before.

When our time was up, we returned home and Sophia was so happy and doing so great. I on the other hand was doing a good job of hiding everything I was going through thanks to the meds the doctors had me on. Each week I returned to therapy which only opened up wounds even more. I learned cigarette smoke was a trigger for me which was why I had fits of rage every time I smelled it. Considering my mother was the only person I was around regularly that smoked, she caught the brunt of that rage often. I tried to tell her it was the cigarettes I hated and that I had ripped complete strangers apart for it and she couldn't understand.

Then during therapy I realized the sense of smell is most closely connected to memory. My attacker reeked of cigarette smoke which was why every time I smelled it, I freaked out. When I realized what I had been doing to my mother all those years by taking out my anger on her, I felt horrible. I cried for days before I had the courage to explain myself to her. She actually appreciated me explaining things because she could then understand it wasn't a personal attack against her. Things improved a little bit as she made an effort not to smoke around me. But eventually she went back to her old habits of smoking anywhere and anytime she wished...which was 24/7. Needless to say, I still got on her case about it, but I tried to choose my words more carefully rather than tear into her as though she was the enemy. At 73 expecting her to change was futile.

Eventually my therapy came to an end and in all honesty, I didn't think it did me a bit of good. In fact, I told the counselor I felt I had gone backwards instead of forwards. I had withdrawn back into my cave and lost all interest in dating.

Revisiting all those horrible memories and the pain that came with them in addition to all of my experiences in law enforcement made me trust people even less. The therapist apologized for making me feel that way, but told me it could take some time for the work I had done to take effect. I politely said, "OK," but I knew where my head was and it wasn't good.

By this time I had new supervisors at work who were much more supportive and understanding of not just me, but everyone on the squad. They understood the value of family and that many of us led separate lives outside of work. That, in itself, was a nice breath of fresh air.

As the months went by and I continued to leave for therapy, I was forced to tell one of my supervisors why I left work early every couple weeks. After telling my story at the VA a few times, I was getting more comfortable talking about it with people I trusted. My corporal was awesome in showing support and did whatever he could to make sure I had what I needed and got the support necessary.

Summertime came and Sophia and I decided to pull the camper to Oklahoma so it would be there when we returned the following Christmas. The purpose was so I could sleep in it with the dogs. Maggie was allergic to dogs so they weren't welcome in the home. Oklahoma winters are brutal and there was no way I was leaving the dogs outside. On the way to Oklahoma, we passed through Mississippi and met with Zach. We played some long overdue games of Scrabble and as usual, he beat me. Somehow I would always beat him on our phones using a scrabble app, but in person, he would smoke me. Needless to say, we had a great evening of catching up and Sophia finally got to meet him which she had been wanting to do for a very long time. Zach and I went for a walk and talked. Despite the fact that we hadn't seen each other in thirteen years, it was as though not a day had gone by. The emotions and chemistry remained, as did the love.

The next morning Sophia and I pulled out early and headed out for a long day on the road to Oklahoma. We passed through parts of the country I had never seen. I never realized how beautiful farmland could be and we would go miles without seeing another car as we passed through southern Arkansas.

Late that night we finally arrived in Oklahoma. As we arrived at Damon's house, he and Maggie met us outside. I could see something was wrong in her eyes as she approached our truck. Apparently they had received a lot of rain in recent days and because it was dark, I accidentally backed over the grass adjacent to the entrance to their driveway. She seemed very concerned about how Damon would feel about that as she said "He's been in a mood and hasn't spoken to me for days." As we backed in, he asked to park the RV where he wanted it. I gladly let him since I was exhausted from driving. It didn't take long for me to figure out, something was wrong on the home-front. You could cut the tension with a knife.

It was late, everyone was tired, so we just went to bed. The next morning, Damon was gone. I met with Maggie who said all he does is get up before dark, go to the ranch and come home late every night. When he was around, he wasn't nice to be with. Something was definitely wrong and I had serious second thoughts about leaving Sophia there when I left. She so badly wanted to get to know her father and asked to spend time there on her own.

Originally we had plans to go to the Oklahoma City Memorial, but Damon never came home that day. I began texting him telling him his daughter just traveled for two days and over 1300 miles to spend time with her father, so where was he? That night he returned and he agreed to take us to the memorial the next morning.

When the next morning came, there was a change of plans. Instead of everyone going to the memorial, it would just be Damon, Sophia and me. Apparently he didn't want to spend the day with Maggie and her boys.

Before we even got out of the neighborhood, Damon turned to me and Sophia and said, "Nothing personal, but I'm not around because I can't stand her kids." That's when it all made sense, but, the look on Sophia's face said it all. It was as though I could read her mind as she thought, "What if he doesn't like me?"

Damon continued to explain that his second wife had boys and he couldn't stand them either. He justified his actions and feelings by saying he "was better having daughters because they were easier." Whether or not that was true, he probably should have kept that between him and other adults. It was the wrong message to send to a daughter he barely knew.

The day was uneventful until we got back to the house. I had planned to leave the next morning to make it home in time to go on a cruise with friends, but I was concerned about making the long drive by myself, so I decided to get a head start so I could make it home in time. As I explained to Sophia it was time for me to leave, I pulled her to the side and I told her I had a bad feeling about leaving her there. I told her I didn't think it was a good time for her dad and Maggie and that maybe she should come home with me until they got things sorted out. Sophia stood there crying for the first time ever that I was leaving her with someone. She never even cried when she went off to preschool or kindergarten. She had never been a clingy kid. She was independent like her mom. It broke my heart to see her like that and I begged her to come with me, but she said "No, I'm never going to get to know him if I don't try." I was proud of her for her strength, but at the same time, my gut instincts were telling me leaving her was a bad idea.

I told Damon I was concerned and that her emotions were not normal for her. I told him to SPEND TIME WITH HER and try to find things to talk about. He told me not to worry, she would be fine.

As I headed out, I made a wrong turn and had to call for directions to the interstate. Damon didn't answer his phone so I called Sophia's. When I heard her voice, I could tell she was crying. I asked her where she was, she said she was in her sister's room. I asked her where her father was and she said he was watching a movie. I suggested she go watch the movie with him and she told me it wasn't the kind of movie I would want her to watch. When I asked her what kind of a movie it was, she said it was just really violent and showed people getting shot and blown up.

I texted Damon and suggested he save the movie until later and told him she wasn't comfortable watching violent movies. I suggested they play a board game and talk. Later I would learn he went to pick up his preacher's daughter to bring her over to play with Sophia. He literally had no idea how to talk to or interact with kids, nor did he want to.

On my way home, I made it as far as Mississippi and met up with Zach. He took some time off work so we could spend some quality time alone. It was an emotional night as we revisited the past and pondered the future. True love never dies and ours hadn't. He wanted to make plans to see me again but I told him I couldn't until he had his divorce papers. I had made a deal with God and I had just broken it after 13 years. I felt guilty for breaking my word, but I loved him and no matter what happened and how many years or miles in between, I couldn't change that.

When we parted, neither of us really knew if that would be the last time we'd be together, but I hoped with all my heart it wouldn't be. I hoped he'd finally find the guts to follow his heart, but only time would tell.

I arrived home and began to pack for my cruise the very next day. I had a lot to process and think about, but I couldn't get my mind off Sophia.

As the week progressed, I took comfort knowing Sophia's aunt, uncle and cousin would be coming from Texas again. The whole family had planned a camping trip for July 4th. I figured things would get better as long as others were around to keep her busy, but as the week went by, I could tell something wasn't right when we texted. She wasn't giving me details, but it was obvious she wasn't happy.

Finally I got home from my cruise and Sophia was due to fly home the next day. That night I received a text from her saying, "Mom I hate him. I'm mad at God and I'm mad at Dad! I want to come home." I immediately called her and told her to go somewhere she could talk privately. She locked herself in the bathroom and just sobbed like I had never heard her before. It broke my heart and I just wanted to get her home that instant. I told her I could understand why she might be mad at her dad, but why was she mad at God. She told me she had prayed her whole life for a dad and this is what she got. She went on to say how she wished she had never met him. She told me he was a horrible person and had nothing nice to say to anyone. Then he said something to her that utterly broke her. He yelled at her for seeking air conditioning when he had everyone outside in 110 degree weather. He told her, "I don't like lazy people and you're lazy!"

While I will be the first person to admit kids can be lazy and don't like to do physical labor, this was different. I have experienced heat in the jungles of Panama and nearest the equator, I have NEVER experienced heat like I did those few days I was in Oklahoma. It was brutal and Sophia wasn't used to it. She overheated easily at home and I had warned him of that. Nonetheless, he was working her and the other boys like slaves in the hottest part of the day. I was furious with him. Sophia went on to tell me how he hadn't spoken to her in four days and she would wait up for him at night just to say goodnight, only to have him walk by her silently as if she wasn't there.

It was all I could do to keep it together. I was so furious, I wanted to kill him. I spoke with Maggie who backed up every word Sophia told me, so I knew she wasn't exaggerating. It was then I decided to tell Maggie that no woman or her children should have to live that way and I didn't understand why she tolerated it because I certainly wouldn't.

Maggie was a devout Christian woman so she told me she was just going to keep praying for him to change. That never happened and a year later, he got his third divorce.

It took a lot of counseling, countless hours of crying and a team of women to get through to Sophia that what her father did was not her fault. I convinced her that God answered her prayers when He thought she was mature enough to handle the type of man her father was. Not every parent is perfect because we are all human and have our own issues. Damon was carrying years of child abuse and the battle scars from war in his head. He was refusing to deal with it or accept that HE was the problem, not everyone else. No one knows if he ever will, but that is on him because God knows he has had ample opportunity and invitations to get help. As for Sophia, she eventually came to the conclusion that her father was a jerk, but she didn't hate him. She also figured out that while she may not be able to have a relationship with her father, she did end up with an awesome big sister who she looked up to and continued to grow closer to over the years.

Twenty-Three

Fighting Stigmas

Florida, 2020

While I managed to keep my diagnosis of PTSD quiet with my employer for fear of losing my job, it seemed like everywhere I turned, I was meeting veterans who, based on their military experience, were fighting their own demons and refusing to get help for fear of the repercussions because of the stigma attached it. Some were parents, coworkers and others were retired cops or firefighters. People don't realize we are everywhere and we have learned to acknowledge each other with a look of compassion and unspoken understanding.

Our country's bravest are suffering in silence because every time there is a mass shooting, the media reports whether or not the assailant is a veteran and if they are, they blame his actions on PTSD. I can assure you, PTSD does not cause people to go on mass shooting sprees. The only person someone with PTSD wants to hurt is themselves. While they might be abrasive with their words or unpleasant to be around, they don't want to commit violence. They are just miserable and in pain and often that misery comes out in a way that frightens others. We tend to have nasty tempers and say things we shouldn't say. Sometimes we just need some peace and quiet to get ourselves together. Often we see things that others don't and we get frustrated because when we see danger, they accuse us of being paranoid. We are hypervigilant and we overreact emotionally. It sucks and we hate it, but that is what our experiences have done to us.

Sometimes PTSD causes medical problems. Gastrointestinal problems are common as are headaches, restless leg syndrome and many other physical ailments. PTSD sucks and unless you know someone has it, you're likely to just think they are an asshole or a bitch. We get it.

What sucks the most is the stigma PTSD carries and when someone says to you, not knowing you have it, "He's dangerous...he has PTSD from the war." I can't count how many times I've heard regular citizens say that not knowing I struggle with it myself. It makes me angry and I feel the need to defend those they are talking about. These people are heroes and they are broken because of what they've seen other horrible human beings do to one another. The negative stigma is the fault of the media and journalists who don't take the time to do real research. All they care about is creating a juicy story without all of the facts.

Have people with PTSD murdered others? Yes, they have, but not BECAUSE they have PTSD. Often they have other mental health problems like schizophrenia or bi-polar disease or they are just simply evil. Often these mental health issues don't surface until someone is in their twenties and after they've joined the military. It is very common for mental health problems to go undiagnosed until they get out of the military and then they begin to act on them. These are the people that fall into conspiracy theories or become angry with the government because they blame their problems on the military rather than something they inherited genetically.

It's heartbreaking and instead of avoiding our heroes who are struggling within, our society needs to understand what they are going through and do more to offer help. Let them know it's OK and that they aren't alone. That kind of support is incredible and in and of itself is a huge help.

While our Veterans Administration is slowly beginning to improve thanks to our current administration, there is still much work to be done as it took generations to get this bad. Many of us believe the VA should only hire veterans so that they understand how to communicate and interact with other veterans. There is nothing worse than being a veteran and trying to discuss things with a civilian that just doesn't understand what we are going through and how we got to this awful place we are in or how to help us get out of it.

One of our biggest successes and accomplishments is all of the amazing programs and people that provide veterans in need of a furry best friend. Our society, employers, companies and everyone else needs to understand that not all disabilities are visible. Let a veteran or first responder with a service dog go wherever they wish without being hassled.

Folks need to stop making accusations that people are using fake service dogs. Are they? Absolutely they are, but instead of giving them the stink eye or insulting them by asking "What's wrong with you that you need a service dog?" simply watch and observe. If the dog is well behaved and not acting out by pulling on a leash, barking or misbehaving, assume they really are a service dog and smile and just say Hello. If you show them you are generally curious to know more about their dog, ask them how they trained them or if they think they could help someone you know who suffers from whatever condition you're thinking of. If they want to talk to you, they will. If they don't, then don't take it personal if they avoid conversation. It is extremely difficult for someone with PTSD to go into crowded areas like malls, concerts or fairs. The anxiety is real and our dogs are a huge help to us. This past November I managed to go Christmas shopping in public for the first time in 30 years, all thanks to my dog, Saint. It was an awesome experience I got to share with my daughter and it was a huge accomplishment for me.

On the other hand, if you see a dog acting up, barking, pulling on their leash, acting like it is more interested in everyone BUT their handler, then assume that dog is a fake. There are two questions that can be legally asked, "Is that a service dog?" and "What service do they provide?" If they answer those questions and the dog is acting up by barking, jumping, pulling or causing a disruption, then legally they can be asked to leave by management.

If businesses and citizens were to educate themselves, then truly deserving veterans and first responders wouldn't be hassled when they are out and about. In addition, even if someone isn't a veteran or first responder, service dogs are now found to be beneficial to those who suffer from seizures and diabetes in addition to other invisible disabilities. So people just need to use their brains and observe before they jump to conclusions or assumptions that someone is a fake.

When talking to a veteran or cop, don't ask questions like "Did you ever have to kill someone" or "What is the worst thing you've seen?" Trust me, these are memories they are trying to forget. They are dark and they are scary. They certainly don't want to discuss them with someone they barely know or won't understand. IF they want to talk, they are going to talk to their brothers or sisters and those who've been there and understand.

I don't care who someone is, from the richest to the poorest and everyone in between. We all have a story and we have all been through tough stuff the longer we've been on this earth. Our "stuff" leaves scars, some bigger and deeper than others. Some might only have one scar, some of us have several. Regardless, we need to be able to accept that of ourselves and be courageous enough to admit we have a problem and ask for help, before our demons rise up to pull us to our graves.

Despite everything I've endured in my life, I have finally gotten to a place that I am comfortable admitting I'm far from perfect. I am broken and I don't know if I will ever be fixed. As I embrace turning the BIG FIVE ZERO, I have finally learned to accept my scars and I have finally learned to focus on all the Blessings I actually have. I have amazing funny stories to tell that built me into the strong woman I am today. They say, "That which does not kill you, makes you stronger," well I'm living proof. I am Blessed to have amazing friends who have been there

COURAGEOUSLY BROKEN ~ 271

for me through the best and worst of times. We don't see each other often and we might go months or sometimes years without talking, but the true sign of friendship is when there are miles and years between, we can all pick up right where we left off as though it was yesterday. I don't know many citizens that can say that, but I promise you, there are countless veterans and first responders that can.

I have an amazing daughter who is now a teenager and while I have days I worry myself sick over her and she makes me feel as though I am losing my mind, I still know that compared to other teenagers, God Blessed me with a daughter that most could only dream of having. I know this because of all the compliments I get on her. Someday when she is grown and on her own, I have no doubt she will be my best friend and I hope I will be hers.

When I look back at many moments in my life that I questioned, "why?" I can look back and see why God put a certain person or event in my life. It is only then it makes sense, if you just look at it from the right perspective. For those that don't, just accept that shit happens and don't let those times hold you down.

Everyone wants a perfect life or a perfect family, but the fact of the matter is, that just doesn't exist. As a cop I've seen more dysfunction in multi-million dollar homes than I have some of the poorest homes or families I've encountered. We are all human and we all put our pants on one leg at a time. None of us is better than anyone else and if we could learn to just treat others the same way we would like to be treated, the world would be a much better place.

Lastly, remember this...if the good people of this world were sheep and the evil people were wolves, remember that the brave military and law enforcement officers are the sheepdogs. The sheepdogs aren't always friendly or cuddly and they might even bark at a sheep if they get out of line, but they are still there to protect them and they will kill the wolves if they threaten the sheep. Remember that one day, the sheepdog will retire or move on to other things, but they will always be willing to lay down their life to protect the sheep if they have to. Let them do their jobs and remember to say thank you. Sometimes good people have to be willing to do violent things so that the innocent can live in peace. Give them the respect they deserve and unless you are willing to run into the gunfire with them, don't question how they do their job, because they aren't judging you on how you do yours.

I don't know what my future holds, none of us do. What I know is that I plan to retire and hang up the uniform as soon as I can. My daughter will be off to college and then I hope to keep checking things off my bucket list. We've got this big beautiful world and while I've been Blessed to see much of it, there is so much I haven't seen. We only get one chance at this thing we call life. Why not experience as much of it as we can? Personally I have found making goals a way to keep putting one foot in front of the other. I know when my nest is empty and it's just me, I'll have my furry best friend to take along on my adventures.

Will I ever find love? Will I ever settle down? I honestly don't know. I've come to accept that some things are just out of my control no matter how much I hate it. If I'm meant to be with someone, then it's in God's hands. Hopefully God will make it painfully obvious to me that HE IS THE ONE if and when I meet the right guy. All I know is the bar is set pretty high and he better be pretty amazing to gain my trust and win my heart.

I mean think about it, I spent my early twenties surrounded by some of the bravest, toughest, most handsome men on the planet who were brilliant and hilariously funny. I fell in love with one and he remains my best friend. Those are some pretty big shoes to fill, so while I am open to love, Mr. Right better really have his shit together because after everything I've been through, I've finally accepted that I deserve it. I may have scars and I may have baggage, but I also have a really big heart and if I'm treated right, I will return the love and respect tenfold. Isn't that the way it should be for everyone? I'd like to think so, but who knows? Time will tell and that is the adventure of life.

In the end, the best advice I have to give based on my roller coaster life, is to never judge someone because of the faults of another. Are there bad cops out there? Sure there are and according to the numbers, it's about .01%. That means that 99.99% are good and choose to face danger because they genuinely want to help people. Sadly, over the years of being cursed at, judged and raked over the coals on television, the passion to go the extra step for people they perceive hate them gets much more difficult with each passing day, yet they still do it. Let them know you support them. A kind word and smile will go a long way, I promise.

Remember, there are bad apples in every bunch. There are men who hurt women, there are women who only see men as a financial benefit and use them while leaving them heartbroken. There are priests who hurt kids and lawyers who make backroom deals with judges. There are doctors who judge patients or cut corners causing harm. The list goes on and on of human beings taking advantage of their positions to hurt others for personal gain, but that doesn't mean that all humans are bad. What I learned in the Navy was that the vast majority of men I served with were great men and that it wouldn't have been right to hold the faults of one against so many. As human beings we need to stop judging or hating an entire gender, race or occupation because of the faults of a few.

I wouldn't be where I am today if it wasn't for those three years in Panama and I learned from the best what it means to "Never Quit." I arrived there broken and they tested me, teaching me to be stronger than what I had endured. Years later they were still there for me when years of trauma caught up to me. I am able to recognize while there are horrible evil men out there, there are plenty more who are courageous warriors ready to lay down their life protecting mine and they will forever be my brothers and I will cherish their friendships until the day I die.

So although I've been through my fair share of pain, I know I'm not alone and that we all have a story to tell. I've learned that if you open your eyes and take the time to really get to know people, you can find plenty of good ones and that brings me hope for the world.

I realize no one can fix me but me. I still struggle to forgive myself for the mistakes I've made in the past. I worry about how many other women that asshole raped and if I could have saved them from the same trauma I've endured by reporting him. With all the therapy I've been through, I'm not sure I will ever stop asking myself that question. I would like to know whatever became of him. Did he do it again? Did he get caught? Was justice ever served? I'd really like to know in the hopes that I might get closure on it.

I hate having to tell someone about my PTSD and why I have it because I don't like to be pitied or judged. I don't want to hear an innocent person tell me they are sorry for what happened to me even if they mean well. I'd much prefer they ask sincere questions if they have them and respect me if I'm not ready to answer them quite yet until I know and trust them better.

Maybe eventually I will meet the right guy who has the patience to bring down my walls and gain my trust. I would like to grow old with someone I love and who loves me despite my battle scars. Only God knows if I will ever find the right guy for me. Until then, I'll just keep doing what I do and that keeps me plenty busy in the meantime.

Twenty-Four

❧

Reflection & Giving Thanks

As the years have gone by and I've met others who've had similar experiences to mine, I realize many are not coping as well as I am. I don't mean that in a judgmental way, because some of them have done better in other areas such as finding love. They've managed to marry and have what looks like a normal life. Meanwhile behind closed doors, they struggle with alcohol, prescription drugs or anger management issues.

I, on the other hand, have not found someone I can settle down with. I've always run back to my "safe place" with Zach. I am well aware there will be many who will judge me for my actions and that is their right if they choose to do so. I don't expect anyone to condone the choices I've made any more than they can understand my reasons why. I've learned not to judge others because as imperfect human beings, we all have our faults.

As I reflect on my choices over the past many years, I realize I was often my own worst enemy and was self sabotaging any chance of finding a healthy relationship. I engaged in long distance relationships and was drawn to men who clearly had commitment issues of their own. When that got old, I let my hypothyroid disease become an excuse to let myself go so I could repel men. I continue to have an uncanny ability to find something wrong with every man I meet who shows me interest. It's a classic example of, "It's not you, it's me" and I often feel bad when they don't believe me even though it's true.

While I have loved and lost, I have succeeded in other areas. I've somehow managed to have a successful career, raise an amazing daughter, own the home I've always wanted and soon I hope to have a comfortable retirement. Hell, I've even been able to buy the car I've dreamed of since I was a little girl. I've traveled to incredible places and I hope to travel to more in the future.

I'm certain I would not be where I am today without the support of amazing people God has put in my path. As much heartache as Zach has brought me through the years, he's always been the one constant who knows what to say when I'm drowning in the darkness and can't pull myself out. He believed in me when I couldn't believe in myself. He's given me advice when my rage and emotion have gotten the best of me and God only knows how much grief that has prevented. Every time I let my anger and emotions take control, I just seem to make things worse. So while many won't understand why I continue to keep him as a part of my life, I see the good in him and the great man he's become. Mostly I love him as my best friend who no matter what, is always there when I need him most. He has and always will be my rock and after 30 years of unconditional love, that's not going to change. Only God knows why things are the way they are and someday I fully expect it to make sense. In the meantime, Zach will remain my bestfriend and soulmate. For anyone who may wonder about his wife...Yes, she knows about me and I've been told she's been given the choice to accept our friendship or not. Either way, neither of us are going to abandon one another regardless of the many miles between us.

God puts people and events in our lives for a reason. I firmly believe I was sent to Panama to meet the people who would change the course of my life forever by teaching me the meaning of "Never Quit." God knew my struggles and the ones yet to come. Zach would become my hero who would someday rescue me from myself. Damon, who against all odds, would eventually become the father of my child. Reagan who would recognize my pain and give me a dog that would become the greatest medicine, bring so much joy, unconditional love and loyalty unlike no other during a time when I needed it most. Raul, the smartest man I've ever known, would provide me with knowledge and wisdom necessary to make so many difficult decisions when it came to parenting and medical decisions. Nichole, the sister I never had would be the shoulder I would cry on and the one I could laugh with over things that only she and I could understand. The life long memories and laughter we share every time we get together always leaves us wanting more. So we bid farewell until next time.

They say it's a small world and it is. After 30 years, a couple from Panama, Nick and Kari, settled down only an hour from me in Florida. They host a reunion that grows bigger each year and there is something so cathartic for all of us when we get together and talk about old times and catch up on each other's families and lives. There are other events hosted by organizations like the Navy SEAL Museum every year in Fort Pierce, Florida. Every year we all meet up, reconnect and enjoy the tremendous camaraderie that never dies. In fact, it only grows stronger with time.

In honor of the people I credit with building me into the woman I am today, I named my dog Saint Michael, after the Patron Saint of Military and Law Enforcement. The story of Saint Michael is profound and it makes sense why he is our Patron Saint.

After God created the heavens and earth, a battle began in heaven. Lucifer, which means the "light-bearer" in Hebrew, was the chief among all the angels. Lucifer rebelled against God because of his envy, pride, and desire to sit upon God's throne. Lucifer did not want to be beneath God or to serve Him. He took a third of the angels to fight with him. An angel named Michael however, was loyal to God and declared he would serve God. In Hebrew, Michael means "who is like unto God." Michael gathered the other two-thirds of the angels and defeated Lucifer and cast him and his supporters out of heaven. Lucifer became known as Satan, which means adversary, and those angels who supported him became devils. As a reward for his loyalty, Michael was made the chief angel. Due to this leadership role, the Church named the Archangel a Saint and considers him to be the "highest general" in God's army of angels.

When I think of Saint Michael, I like to refer to him as the *badass angel* because of his ability to defeat Satan. It makes perfect sense why he is our Patron Saint. Sometimes good people are forced to do bad things in order to protect God's children. Our military fights those who are evil in foreign lands to protect those who are suppressed in their country in addition to stopping the same evil people from coming to our homeland and harming us. Police have the job of protecting the good from the evil that walks among us here at home. Under our tough exteriors, we enjoy doing good and helping others, but it comes with a cost sometimes and that leaves scars.

Many people question God and why he allows bad things to happen to good people. I am extraordinarily familiar with those feelings. What I've figured out is that God doesn't allow those things to happen. As far back as the beginning of time there has always been a battle between good and evil. The bad things that happen in our life are not because of God, they are because of Satan trying to turn us away from God in an effort to build his army in the hopes of defeating God when our time on earth is done. It is up to us to recognize this when that day comes and to remember that if we trust in God and believe he is there to support us through our difficult days, we will be able accept that everything happens for a reason. We need to believe that while these tragedies won't kill us, they will make us stronger. Maybe our future holds something where we will need to call on that strength or maybe we can use our strength to help someone else struggling through similar tragedy.

Every single human being on earth will go through life with experiences that will leave them with scars. Some will be deeper than others. How they choose to wear their scars is up to them. Medal of Honor recipient Sammy Lee Davis wrote a book entitled, *You don't lose until you quit trying*. My former skipper from Panama told me about him and his book. That title speaks volumes and I plan to read it in the near future. What he says is true, we WILL get knocked down in our lives, we WILL get hurt and we WILL get scars. Some scars will be visible, some won't, but they will all leave a mark nonetheless. It is what we choose to do despite all of this that counts and as long as we don't stop getting up, putting one foot in front of the other and fight for the life we deserve, we can't lose.

I know many will read my religious views and dismiss my theories. That is their choice. I, myself, have been there and I understand the doubt and resentment pain can cause someone to feel. I'm fortunate that I no longer have those feelings of doubt thanks to the people and experiences I've had over the years. I've been given the gift of hindsight. I will be the first to agree that although I am Catholic and choose to raise my daughter as Catholic, I don't necessarily agree with everything the Catholic Church teaches. I know every man and woman on this earth is human and to err is human. Our Church Leaders, regardless of their denomination, make mistakes and don't always do the right thing. Churches are full of hypocrites, but that doesn't matter because what matters is the relationship we, as individuals, have with God and how we live our lives in an effort to support good over evil.

I don't know what my future has in store for me, but I hold on to the hope that comes with these beliefs. I hope by sharing my experiences, I can help others who have had similar traumas who have also struggled with sleepless nights and dark days. Always remember, a permanent solution to a temporary problem is never a good idea. Talk to someone, call a battle buddy. Trust that none of us are alone in fighting our demons. Together, like Saint Michael, we are an army and we can beat the demons who haunt us. Don't be afraid to ask someone, even a stranger, if they need to talk. People are put in our paths for a reason and one day, that person could be you.

Being a hero doesn't always involve courage or bravery. Sometimes it's as simple as lending an ear to listen or making a small gesture that means the world to someone. So, whenever the opportunity arises, be someone's hero. You never know who's life you might change or the possibilities that may come.

Lastly, remember this, if you've suffered from the battle scars of life, causing you to feel broken from time to time, you are never truly broken as long as you are fighting to survive. Be that as it may, if you are going to identify with being broken, then be Courageously Broken, but never defeated.

"She was brave and strong and broken all at once."
~ Dr Anna Funder

There is Help

22 A DAY... 22 Veterans commit suicide every single day. Despite all that has been done to bring this number down, it remains. The question is Why? After all, enough awareness has been created that it now has its own "Awareness Day". Perhaps it is the type of awareness that we need to look at? I know my veteran and first responder friends will understand this, because many of us live this life. Sadly though, much of the public doesn't truly understand and I believe that is because of the media.

When you hear the term PTSD, don't believe everything you hear or buy into the fears so many believe. I know COUNTLESS people who are everyday heroes who walk among us with PTSD and you would never even know it. They are Veterans, Cops and Firefighters. They could be children or spouses who've been abused or sexually assaulted.

PTSD is real and it can affect anyone, even the strongest. If addressed properly, it doesn't have to break you. Dealing with trauma takes time and shouldn't be done alone. If you happen to get a counselor and you don't like them, find another one. Some are awful and some are amazing.

If you or someone you know suffers from PTSD and has been struggling, don't be afraid to reach out and talk to them. You don't have to be a counselor, you just have to be a friend and a good listener. Encourage them to get professional help and help them find resources sooner rather than later. The last thing a veteran or first responder wants is to see their worst fear imagined which is being hauled off in a pair of handcuffs because someone had to call the police. The last thing they need is to suffer the humiliation that comes with that experience. Find help before it gets to that point.

If you have questions or need help finding resources for someone you care about, contact SAMHSA's website and search for PTSD support or go to their website https://store.samhsa.gov. You can also reach them at 800-662-HELP (4357)

Acknowledgements

To my daughter, without you, I would have never become the kind of woman I needed to be. You gave me purpose and saved me from myself. I hope and pray that I have done everything I could to be the mother you so much deserve. From the moment you were born, you demonstrated strength, determination and inspired so many. You are truly beautiful both inside and out and I look forward to seeing you grow into the woman you were destined to be. Please know I will always be cheering you on with all my heart.

To my mother I can only say Thank You for finding the courage and strength so many can't find. No matter what, you have always been there to support me, even when I had the craziest of ideas. You've clearly beamed with pride, even when I made it challenging for you. With all my heart, I love you and thank you for being such a strong mom and adoring grandma.

To my brothers and sisters in blue, stay safe and remember to make your families your priority. Don't let the job define you because when the day comes to walk away, it will be your family that counts. You will never wish you promoted higher, worked more or took less vacations. Live life and remember, being a cop is not WHO you are, it does not define you, it is just a job, don't let it consume you. Always do your best, but more importantly, go home everyday, even if it means catching the bad guy another day.

Last but not least, to my military family. Thank you isn't nearly enough. Thank you for the amazing memories and laughter. Thank you for the holidays we've spent together which made missing our real families not so bad. Thank you for the friendships that have lasted a lifetime and for lending an ear to listen when it mattered most. There are just too many of you to list, but I know you know who you are. Until we all meet again for the love and laughter, please know I'll always have your six.

Love Always,
Donna

The Best Things in Life are Worth the Wait

www.ingramcontent.com/pod-product-compliance
Lightning Source LLC
Chambersburg PA
CBHW072338090426
42741CB00012B/2840